Emergence, Complexity and Computation

Volume 37

The Emergence, Complexity and Computation (ECC) series publishes new developments, advancements and selected topics in the fields of complexity, computation and emergence. The series focuses on all aspects of reality-based computation approaches from an interdisciplinary point of view especially from applied sciences, biology, physics, or chemistry. It presents new ideas and interdisciplinary insight on the mutual intersection of subareas of computation, complexity and emergence and its impact and limits to any computing based on physical limits (thermodynamic and quantum limits, Bremermann's limit, Seth Lloyd limits…) as well as algorithmic limits (Gödel's proof and its impact on calculation, algorithmic complexity, the Chaitin's Omega number and Kolmogorov complexity, non-traditional calculations like Turing machine process and its consequences,…) and limitations arising in artificial intelligence. The topics are (but not limited to) membrane computing, DNA computing, immune computing, quantum computing, swarm computing, analogic computing, chaos computing and computing on the edge of chaos, computational aspects of dynamics of complex systems (systems with self-organization, multiagent systems, cellular automata, artificial life,…), emergence of complex systems and its computational aspects, and agent based computation. The main aim of this series is to discuss the above mentioned topics from an interdisciplinary point of view and present new ideas coming from mutual intersection of classical as well as modern methods of computation. Within the scope of the series are monographs, lecture notes, selected contributions from specialized conferences and workshops, special contribution from international experts.

More information about this series at http://www.springer.com/series/10624

Larry Bull

The Evolution of Complexity

Simple Simulations of Major Innovations

Springer

Larry Bull
University of the West of England
Department of Computer Science
and Creative Technologies
Bristol, UK

ISSN 2194-7287 ISSN 2194-7295 (electronic)
Emergence, Complexity and Computation
ISBN 978-3-030-40729-2 ISBN 978-3-030-40730-8 (eBook)
https://doi.org/10.1007/978-3-030-40730-8

This Springer imprint is published by the registered company Springer Nature Switzerland AG
The registered company address is: Gewerbestrasse 11, 6330 Cham, Switzerland

For my girls.

Preface

As a computer scientist who has spent nearly thirty years drawing upon the natural world for inspiration to make machines do useful things, my research with evolution has been somewhat separated from squishy biology. Instead of DNA, I have used evolution with data and/or instructions for computers to design such things as models of Olympic athletes, arrays of interacting wind turbines, nanoparticles for cancer tumour treatment, computers made from chemical reactions, etc. That is, evolution as a powerful search tool by which to negotiate complexity. Alongside that work, I have always turned such abstracted evolution back onto the natural phenomena to help gain insight of the underlying dynamics and emergence/benefits of each. This book brings together much of that work—both old and new—to explore a number of the key increases in complexity seen in the natural world. Whilst any increase in complexity has certainly not been inevitable, they have occurred and this book seeks to explain each of them purely in terms of the features of fitness landscapes.

I have been fortunate enough to discuss this work with so many people over the many years that to attempt to name them all here would be folly. But I would like to thank the Editors for publishing me in their series.

Bristol, UK Larry Bull

Contents

1 Introduction ... 1
1.1 Evolutionary Innovations ... 1
1.2 The Baldwin Effect ... 3
References ... 4

2 Genomes ... 5
2.1 The NK Model: Asexual Haploid Evolution ... 5
2.2 Genome Growth in the NK Model ... 7
2.3 Discussion ... 15
References ... 15

3 Symbiosis ... 17
3.1 The NKCS Model ... 18
3.2 Endosymbiosis in the NKCS Model ... 20
3.3 Horizontal Gene Transfer in Hereditary Endosymbiosis ... 25
3.4 Discussion ... 27
References ... 29

4 Sex ... 31
4.1 The Baldwin Effect in the NK Model ... 32
4.2 Evolution of the Haploid-Diploid Cycle: The Baldwin Effect ... 33
4.3 Two-Step Meiosis and Recombination: Altering the Amount of Learning ... 38
4.4 Genome Growth in Sexual Diploids ... 40
4.5 Coevolving Sexual Diploids ... 43
4.6 Discussion ... 46
References ... 48

5 Chromosomes 51
5.1 Chromosome Number 51
5.2 Sex Chromosomes 54
5.3 Dominance 57
5.4 Discussion 57
References 61

6 Multicellularity 63
6.1 Multicellularity in the NK Model 63
6.2 Functional Differentiation and Simple Epigenetic Control 66
6.3 Eusociality: Haplodiploid Multicellularity 68
6.4 Discussion 71
References 72

7 Conclusion 73
References 74

Appendix: Regulation 75

Index 87

Chapter 1
Introduction

> Complexity is hard to define or to measure, but there is surely some sense in which elephants and oak trees are more complex than bacteria, and bacteria than the first replicating molecules. [7, p. 3]

Life on earth emerged around 4 billion years ago and its complexity has been increasing ever since. This book seeks to explore the conditions under which natural selection [3] would favour some of the key mechanisms by which those increases in complexity have come about, using simple models of evolution on abstract fitness landscapes. Wright [11] was perhaps the first to view natural evolution as a process of adaptation through a multidimensional space of fitness peaks and troughs (Fig. 1.1, top). Turing [10] would later highlight the potential universality of that view when considering ways to design intelligent computers: evolution as a general search process. Whilst no simple correlation between the amount of DNA in a given organism and its perceived complexity exists—lilies have more DNA than humans, for example—it is clear that the two are interrelated. Five ways in which an increase in the amount of DNA may occur are explored here.

1.1 Evolutionary Innovations

The following sources of evolutionary innovation are considered:

Genomes Once formed, increases in the amount of raw material available to evolution in the genome is one way through which new functionalities may emerge, potentially resulting in a novel protein or new regulatory control.

Symbiosis The bringing together of closely interacting organisms such that a new level of functionality is realised by one or more of the partners is ubiquitous and was key to the evolution of eukaryotes around 1.8 billion years ago.

L. Bull, *The Evolution of Complexity*, Emergence, Complexity and Computation 37,
https://doi.org/10.1007/978-3-030-40730-8_1

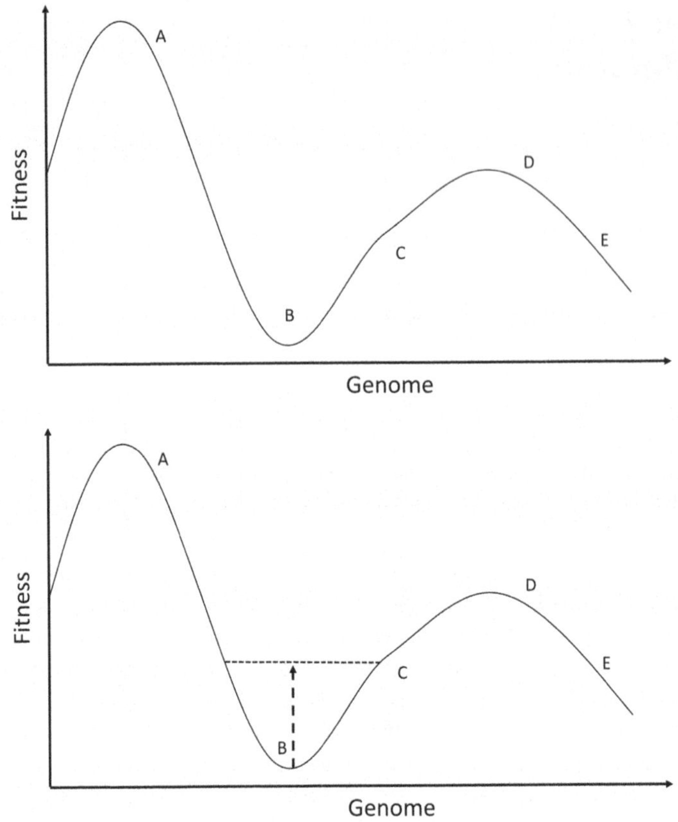

Fig. 1.1 Showing how the shape of a fitness landscape can be altered through a simple learning process. An evolving population at points C-E is unlikely to move to point A due to the deep fitness valley B (top). However, if an individual(s) at point B is able to learn, e.g., to the point C, the effective fitness of the valley can be increased and hence the probability of the population reaching point A is increased (bottom)

Sex Sex involves the bringing together of two or more genomes of the same type to form a more complex cell/organism in comparison to bacteria: syngamy and meiosis under a haploid-diploid lifecycle underpins the evolution of eukaryotes.

Chromosomes Eukaryotes have their genomes arranged in distinct units and the varying of their number is another primary mechanism for increasing complexity, where functional divergence may subsequently follow.

Multicellularity Organisms consisting of two or more copies of their basic reproducing unit opened a significant route to varying complexity around 1.5 billion years ago. Indeed, multicellularity appears to have emerged multiple times. Moreover, eusocial insect colonies can be seen to share characteristics with multicellular organisms which is also considered here.

It can be noted that three of these innovations are also so-called major transitions in evolution wherein independent replicators form new higher level units that can only replicate together thereafter: symbiosis, sex and multicellularity [7] (although, sex is disqualified in [9]). Another process is highlighted as being fundamental to two of those transitions, namely sex and multicellularity—the Baldwin effect [1, 6, 8].

1.2 The Baldwin Effect

The Baldwin effect is here defined as the existence of phenotypic plasticity that enables an organism to exhibit a different (better) fitness than its genome directly represents. Over time, as evolution is guided towards such regions under selection, higher fitness alleles/genomes which rely less upon the phenotypic plasticity can be discovered and become assimilated into the population.

Hinton and Nowlan [5] were the first to investigate the Baldwin effect, showing that enabling genetically specified artificial neural networks to alter inter-neuron connections randomly during their lifetime meant a simulated evolutionary process was able to find an isolated optimum in the fitness landscape, something the system without learning struggled to achieve. That is, the ability to learn "smoothed" the fitness landscape into a unimodal hill/peak. They also found that over time more and more correct connections became genetically specified and hence less and less random learning was necessary; the evolutionary process was guided toward the optimum by the learning process. Their finding was generalized in [2] where it was shown how the most beneficial frequency and amount of learning varies with the ruggedness of the underlying fitness landscape.

Figure 1.1 (bottom) shows a simple example of how this process of moving from the genetically specified position in the landscape can be beneficial by enabling the smoothing of fitness valleys, thereby decreasing the tendency for a population to become trapped on local optima. Learning—phenotypic plasticity—can be achieved in many ways. Alongside neural processing, the Baldwin effect has been connected to other aspects of organisms, such as the immune system [4]. It is suggested here that the bringing together of two or more replicators of a given type into a single organism/entity can be viewed as a rudimentary form of learning, assuming the resulting fitness is a composite, e.g., the numerical average of the partners. For example, in Fig. 1.1, if an individual at point B was combined with an individual at point D, the fitness valley would be greatly diminished since the resulting fitness of (B + D)/2 would be roughly equal to that at point C. Note this may cause the apparent fitness of point D to drop significantly, e.g., if no other purely D individuals exist.

The rest of this book explores the five sources of innovation identified.

References

1. Baldwin, J.M.: A new factor in evolution. Am. Nat. **30**, 441–451 (1896)
2. Bull, L.: On the Baldwin effect. Artif. Life **5**(3), 241–246 (1999)
3. Darwin, C.: On the Origin of Species by Means of Natural Selection. John Murray, London (1859)
4. Hightower, R., Forrest, S., Perelson, A.S.: The Baldwin effect in the immune system: learning by somatic mutation. In: Belew, R.K., Mitchell, M. (eds.) Adaptive Individuals in Evolving Populations, pp. 159–167. Addison-Wesley, Redwood City, CA (1996)
5. Hinton, G.E., Nowlan, S.J.: How learning can guide evolution. Complex Syst. **1**, 495–502 (1987)
6. Lloyd-Morgan, C.: On modification and variation. Science **4**, 733–740 (1896)
7. Maynard Smith, J., Szathmary, E.: The Major Transitions in Evolution. WH Freeman, Oxford (1995)
8. Osborn, H.F.: Ontogenic and phylogenic variation. Science **4**, 786–789 (1896)
9. Szathmary, E.: Toward major evolutionary transitions theory 2.0. PNAS **112**(33), 10104–10111 (2015)
10. Turing, A.: Intelligent machinery. In: Evans, C.R., Robertson, A. (eds.) Key Papers: Cybernetics, pp. 91–102. Butterworths (1968)
11. Wright, S.: The roles of mutation, inbreeding, crossbreeding and selection in evolution. In: Proceedings of the Sixth International Congress on Genetics, vol. 1, no. 8, pp. 355–366 (1932)

Chapter 2
Genomes

Genome length is one of the degrees of freedom exploited by evolution in the variation of organism complexity and this can be seen to have increased over time in some lineages. Metazoan morphological complexity is known to be correlated with genome length, for example [3]. This chapter explores the effects of fitness landscape ruggedness upon the evolution of genome length in asexual haploid organisms. Hence cells/organisms with genomes of linked DNA are assumed to have already emerged (e.g., see [8]).

Novel sequences of DNA can originate through a variety of mechanisms including retrotransposons, horizontal gene transfers, during recombination events, whole genome duplications, etc. A novel sequence may have no immediate function and be subsequently lost/selected due to a deleterious/beneficial mutation (e.g., see [9]), may beneficially/detrimentally alter dosage (e.g., see [10]), enable the subsequent specialisation of a duplicated function (e.g., see [5]), etc. In this chapter the process of novel DNA sequence creation is simplified such that a given number of random genes are added to an existing genome and immediately assigned random contributions to the organism's fitness function. This is explored within the well-known NK model [7] of fitness landscapes where size and ruggedness can be systematically altered. Results suggest that landscape ruggedness, the length of the new sequence with respect to that of the original genome, the presence of gene deletion, and the existence and type of non-stationarity in the fitness landscape can all affect the evolution of genome length. Significantly, increases in genome length are seen across the parameter space of the model.

2.1 The NK Model: Asexual Haploid Evolution

Kauffman and Levin [7] introduced the NK model to allow the systematic study of various aspects of fitness landscapes (see [6] for an overview). In the standard model, the features of the fitness landscapes are specified by two parameters: N, the

L. Bull, *The Evolution of Complexity*, Emergence, Complexity and Computation 37,
https://doi.org/10.1007/978-3-030-40730-8_2

length of the genome; and K, the number of genes that has an effect on the fitness contribution of each (binary) gene. Thus increasing K with respect to N increases the epistatic linkage, increasing the ruggedness of the fitness landscape. The increase in epistasis increases the number of optima, increases the steepness of their sides, and decreases their correlation. The model assumes all intragenome interactions are so complex that it is only appropriate to assign random values to their effects on fitness. Therefore for each of the possible K interactions a table of $2^{(K+1)}$ fitnesses is created for each gene with all entries in the range 0.0–1.0, such that there is one fitness for each combination of traits (Fig. 2.1). The fitness contribution of each gene is found from its table. These fitnesses are then summed and normalized by N to give the selective fitness of the total genome. Thus, whilst simple, the NK model can be seen to capture the fundamental characteristics of genomes and hence it is used throughout this book. However, it is relatively easy to extend the basic model to explore aspects of regulatory control in more detail, an example of which is included as an Appendix to demonstrate.

Kauffman [6] used a mutation-based hill-climbing algorithm, where the single point in the fitness space is said to represent a converged species, to examine the properties and evolutionary dynamics of the NK model. That is, the population is of size one and a species evolves by making a random change to one randomly chosen gene per generation. The "population" is said to move to the genetic configuration of the mutated individual if its fitness is greater than the fitness of the current individual; the rate of supply of mutants is seen as slow compared to the actions of selection. Ties are broken at random. Figure 2.2 shows example results. All results reported in this chapter are the average of 10 runs (random start points) on each of 10 NK functions, i.e., 100 runs, for 20,000 generations. Here $0 \leq K \leq 15$, for $N = 20$ and $N = 100$.

Figure 2.2 shows examples of the general properties of adaptation on such rugged fitness landscapes identified by Kauffman (e.g., [6]), including a "complexity catastrophe" as $K \rightarrow N$. When $K = 0$ all genes make an independent contribution to the overall fitness and, since fitness values are drawn at random between 0.0 and 1.0, order statistics show the average value of the fit allele should be 0.66. Hence a

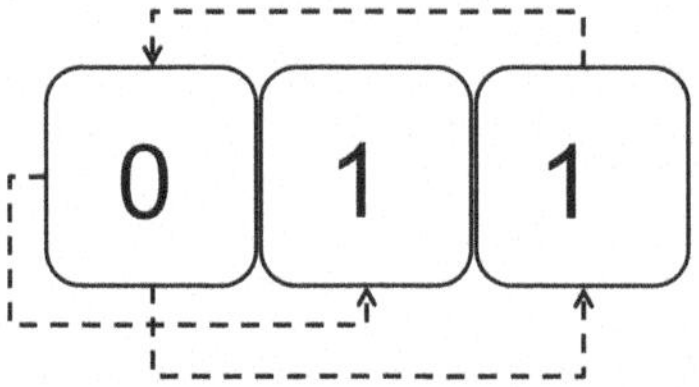

K_0 K_1	Fitness
0 0	0.76
0 1	0.23
1 0	**0.56**
1 1	0.98

Fig. 2.1 An example NK model ($N = 3$, $K = 1$) showing how the fitness contribution of each gene depends on K random genes (left). Therefore there are $2^{(K+1)}$ possible allele combinations per gene, each of which is assigned a random fitness. Each gene of the genome has such a table created for it (right, centre gene shown). Total fitness is the normalized sum of these values

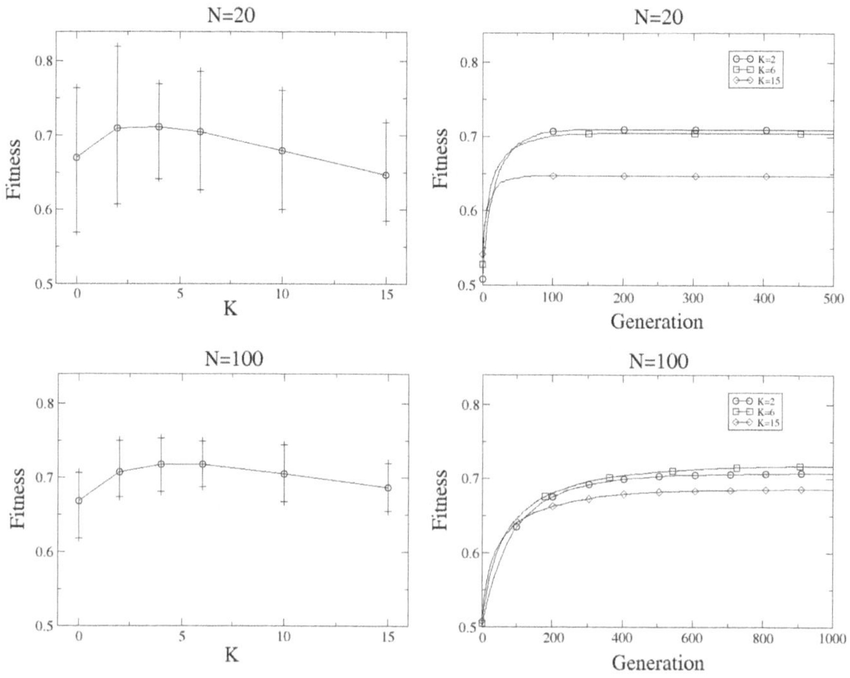

Fig. 2.2 Showing typical behaviour and the fitness reached after 20,000 generations on landscapes of varying ruggedness (K) and length (N). Error bars show min and max values

single, global optimum exists in the landscape of fitness 0.66, regardless of the value of N. At low levels of K ($0 < K < 8$), the landscape buckles up and becomes more rugged, with an increasing number of peaks at higher fitness levels, regardless of N. Thereafter the increasing complexity of constraints between genes means the height of peaks typically found begin to fall as K increases relative to N: for large N, the central limit theorem suggests reachable optima will have a mean fitness of 0.5 as $K \rightarrow N$. Figure 2.2 shows how the optima found when $K > 6$ are significantly lower for $N = 20$ compared to those for $N = 100$ (T-test, $p < 0.05$).

2.2 Genome Growth in the NK Model

To enable the length of genomes in the NK model to vary under evolution, i.e., to vary to N', mutation is here expanded such that either a randomly chosen gene allele is altered as before or a number (G) of randomly created genes are added to the right-hand end of the existing genome. In the latter case, the first connection of a randomly chosen gene in the existing genome is assigned to each newly added gene (when $K \geq 1$). The new genes have K randomly assigned connections into the whole genome, as before. In this way the new genes both affect and are affected by the existing genes.

Figure 2.3 shows results from the simplest case of $G = 1$. With a starting length of $N = 20$, the fitness reached increases for $K > 10$ compared to the traditional case of $G = 0$ above (T-test, $p < 0.05$). This is because the genomes typically increase in length by around 3 genes for all K, i.e., $N' \approx 23$, which is sufficient to avert the onset of the complexity catastrophe at the highest K. There is a lot of variance in the genome lengths which emerge but slightly less growth is seen on average for $K < 4$. When $N = 100$ no significant change in fitness is seen for all K compared to $G = 0$ (T-test, $p \geq 0.05$) but the typical amount of growth seen increases with K, where twice or more growth is seen for $K > 4$ compared to the equivalent case with $N = 20$.

As noted above, varying the degree of ruggedness varies the typical height and number of optima in a landscape. The more peaks of low fitness there are, the longer into the evolutionary search it is likely that a randomly added gene can make a beneficial contribution to fitness. Conversely, evolution can be anticipated to move through a series of relatively high fitness levels on correlated fitness landscapes in all but the earliest stages since fewer, higher peaks exist in the global space. Figure 2.4 shows the generation at which evolution finds an optimum for various K, both with and

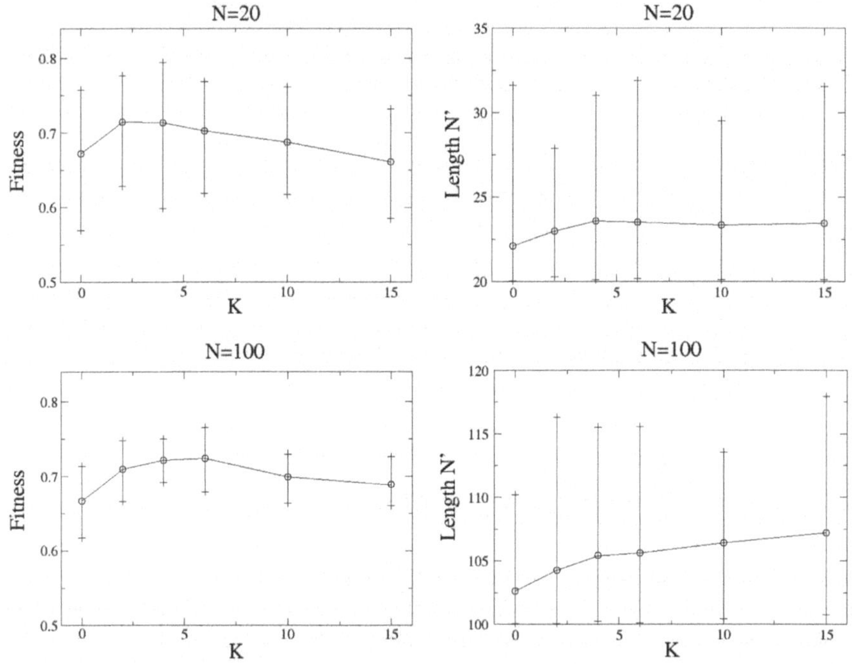

Fig. 2.3 Showing the fitness and length reached after 20,000 generations on landscapes of varying ruggedness (K) where the initial length (N) can increase by one gene under mutation ($G = 1$)

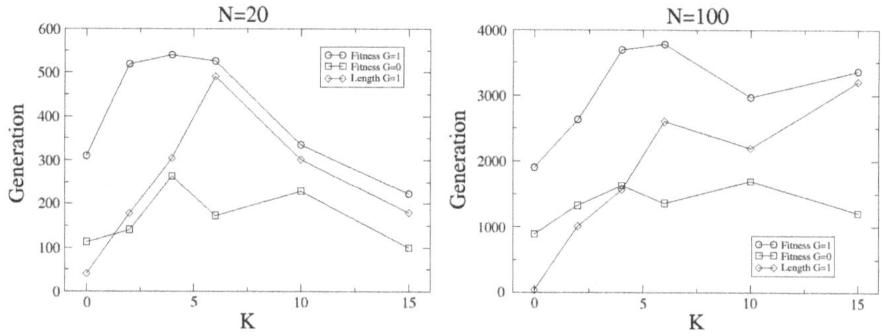

Fig. 2.4 Showing mean walk length to an optimum for a given N and K, $G = 1$. Error bars not shown for clarity

without growth. It also shows the generation at which genome lengths stop changing. As can be seen, for all K, the addition of the genome length varying process means evolution continues for longer before finding an optimum—the dimensionality of the fitness landscape has increased. Moreover, it can also be seen that genome length stops increasing earlier for low K, continuing into the later stages of evolution as $K \rightarrow N$. Figure 2.5 shows the generation at which the first few new genes are typically accepted into the genome for various K. As can be seen, the waiting time does not vary significantly with K, i.e., landscape ruggedness, for the first two genes but begins to vary thereafter.

Figure 2.6 shows the effect of increasing the size of the novel random sequence added, with $G = 20$. When $N = 20$, fitness is greater than with $G = 1$ when $K > 4$ (T-test, $p < 0.05$) as the complexity catastrophe is further averted due to the significant increase in genome length for all K. When $N = 100$, the fitness reached is the same as when $G = 1$ for all K with significantly longer genomes emerging.

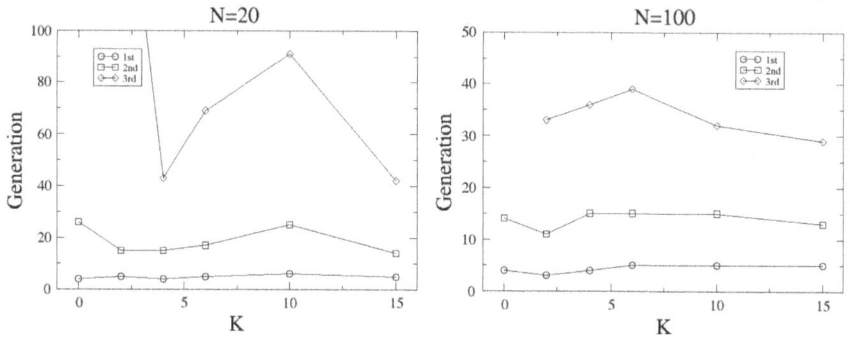

Fig. 2.5 Showing how the waiting time for each increase in length for a given N and K increases, $G = 1$. Note $N = 20$ $K = 2$ typically accepts its third gene at generation 179. Third genes are not added on average for $K = 0$

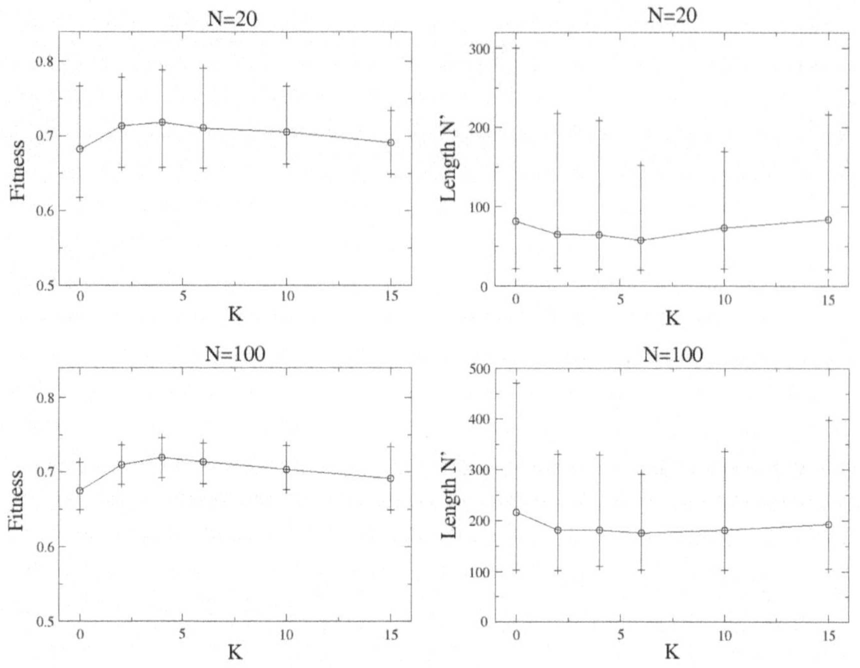

Fig. 2.6 Showing the fitness and length reached after 20,000 generations on landscapes of varying ruggedness (K) where the initial length (N) can increase by twenty genes under mutation

Figures 2.7 and 2.8 show the underlying dynamics of the evolutionary process in each case. As can be seen, the increased amount of growth per addition means evolution continues for much longer than with $G = 1$ (Fig. 2.4), particularly when $N = 20$. The general dynamics are not significantly different when $N = 100$ but the relative increase in size of G with respect to N when $N = 20$ is seemingly more disruptive. Figure 2.8 shows how the first novel sequences are accepted at a similar

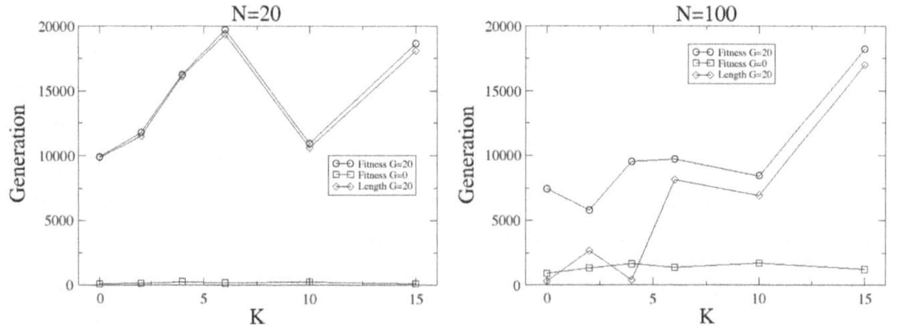

Fig. 2.7 Showing how mean walk length to an optimum for a given N and K increases with growth, $G = 20$

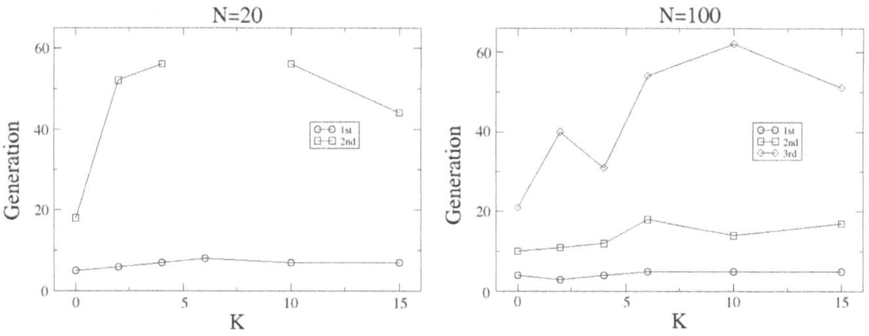

Fig. 2.8 Showing how the waiting time for each increase in length for a given N and K increases, $G = 20$

generation as when $G = 1$, regardless of N and K. The second sequence is accepted at a similar stage when $N = 100$ but at later stages for $N = 20$, if at all. That is, the larger amount of growth per addition event can potentially maintain the conditions for more subsequent growth for longer; each increase in the dimensionality of a fitness landscape supplies a number of sub-optimal gene values, thereby maintaining sub-optimal fitness levels which in turn may aid the likelihood of a new random sequence being accepted.

Using a similarly extended version of the NK model, and $N = 16$, $K = 2$ (only), Harvey [4] showed that gradual growth through small increases in genome length were sustainable whereas larger increases per growth event were not. This is explained as being due to the fact that a degree of correlation between the smaller fitness landscape and the larger one must be maintained; a fit solution in the former space must achieve a suitable level of fitness in the latter to survive into succeeding generations. Kauffman and Levin [7] discussed this general concept with respect to fixed-size NK landscapes and varying mutation step sizes therein. They showed how for long jump adaptations, i.e., mutation steps of a size which go beyond the correlation length of a given fitness landscape, the time taken to find fitter variants doubles per generation. Harvey [4] draws a direct analogy between the size of the novel sequence being added (G) and the length of a jump in a traditional landscape; the larger G, the less correlated the two landscapes. He similarly points out that growth is more likely early in evolution before optima are climbed and it is after that the degree of correlation begins to take more effect. It is here suggested that his larger G did not prove successful primarily due to a change in the standard NK fitness function used to include length, and hence change in the selection pressure, which reduced the size of the window of opportunity for larger increases in length to emerge; results here show larger G increases of genome length are sustainable, for all K, i.e., regardless of the underlying correlation of the landscape. However, it can be seen in Fig. 2.8 that when $N = 20$ a second successful adoption of a novel sequence does not typically emerge when $4 < K < 10$ and a third adoption is rare (contrast with $G = 1$ or $N = 100$ and $G = 20$). It may be that the degree of correlation between the original and new space is playing a role, although growth occurs for $K = 2$ and $K = 4$. It might be expected, although not explored here, that since genome lengths vary for almost as long as fitness levels when a larger G is used, smaller amounts of growth would be accepted more readily during this time. That is, occasional large changes interspersed with many smaller changes, i.e., a dynamic value for G, would provide even more growth.

A common outcome for an added novel sequence is removal, through mutational inactivation, fractionation after whole genome duplications, etc. Figure 2.9 shows example results from adding a process of deletion to balance that of growth described above. That is, an offspring can experience a gene allele mutation or genome length mutation with equal probability, as before. However, in the latter case, there is an equal probability that the last G genes added are removed as for G new genes to be added. Fitness is unaffected for all N and K (not shown) but there is less growth for $K > 0$. The same was also the case for $G = 20$ (not shown).

Given the reduction, the sensitivity of the growth process to the relative rate of deletion to addition has also been explored. Figure 2.10 shows examples of the evolved lengths over the range of no deletion to an equal amount, with $G = 1$. As can be seen, the rate of decline in genome length is roughly proportional to the rate of increase in the probability of deletion.

It was noted above how the addition of new genes typically increases the walk length to an optimum as the population finds its progress "reset" within the higher dimension fitness landscape; new routes to optima in a bigger space become available on each novel sequence addition. Another source of such progress disruption is a change in the fitness landscape. That is, the movement of optima can also cause current gene configurations to become sub-optimal, increasing the likelihood of novel sequences being able to make a positive contribution to fitness. Figure 2.11 shows an example case of the effect on fitness and genome length when the whole fitness landscape is randomly recreated for the given K, i.e., each of the entries in the lookup table of each of the N' genes is assigned a new random value in the range 0.0–1.0, after 10,000 generations. As can be seen, there is a significant drop in the fitness

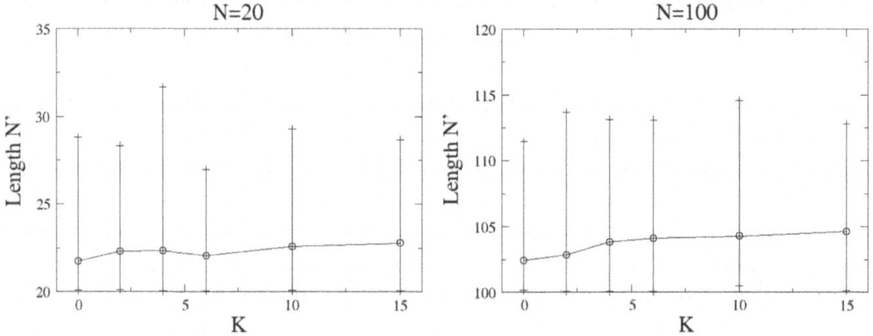

Fig. 2.9 Showing the lengths reached after 20,000 generations on landscapes of varying ruggedness (K) where the initial length (N) can increase or decrease by one gene under mutation

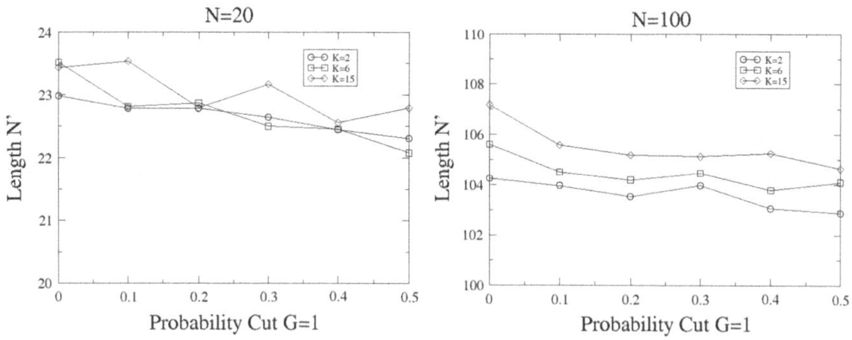

Fig. 2.10 Showing the effect of varying the probability of decreasing the number of genes for different N and K

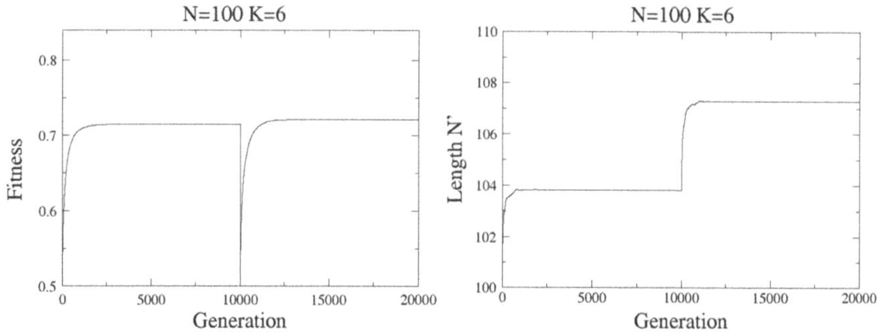

Fig. 2.11 Typical behaviour when the whole fitness landscape changes randomly at 10,000 generations, $G = 1$

level at the point of change before it recovers to a similar level achieved before the change. The effect on genome length is to cause a similar level of growth as from the original length N before the change. Note this variation of the model contains an equal probability of deletion as described above (Fig. 2.9). Growth in response to a change(s) in the fitness landscape has been described in similar models (e.g., [1, 2]).

Figure 2.12 shows the changes in length observed for various N and K combinations after 20,000 generations compared to the lengths evolved after 10,000 generations when the change occurs. Note fitness levels return to around those seen at the point of change in all cases (not shown). The top row shows results from randomly recreating the whole fitness landscape for various G. As can be seen, akin to the example in Fig. 2.11, the effect on genome length is to cause an increase after the change occurs. When $N = 20$ the typical response is to add around G genes, slightly less on average for $G = 100$. When $N = 100$ the typical response is to add 2G or more genes, less on average for $G = 100$. The bottom row shows results from randomly recreating the fitness landscape component of the newly added genes, i.e., the fitness tables of genes N to N' only, at the point of change. When $N = 20$ genes are added in

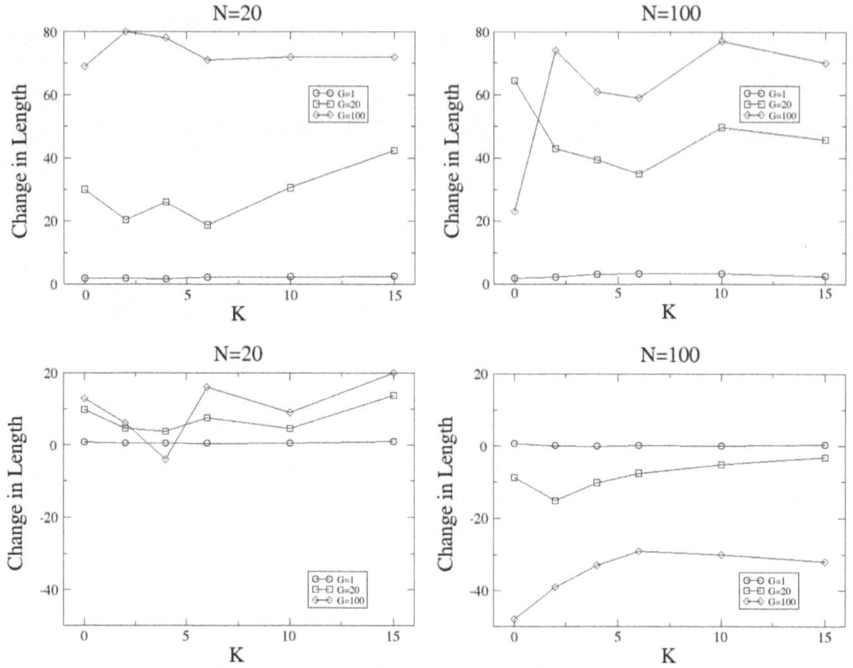

Fig. 2.12 Showing the difference in length between that reached at 10,000 generations when a change in the fitness landscape is introduced and the length subsequently reached after 20,000 generations, for various *G*, *K* and *N*. The top row shows the case where the whole fitness landscape is randomly recreated at generation 10,000 and the bottom row the case where only the entries beyond the initial *N* genes are randomly recreated

the response to the change, although less than *G* genes. In contrast, when $N = 100$, the underlying trend of the response is to *delete* genes. The number of genes added over the first 10,000 generations subsequently removed over the next 10,000 generations typically declines with *K*. As noted above, the number of successful sequence additions is typically higher when $N = 100$, regardless of *G*, and it therefore seems it is more effective to remove some of them to regain a relatively high fitness level. It can be noted that without the probability of deletion, genes are added in all cases, less for the smaller change case N'–N (not shown).

As noted in the introduction, it is here assumed that a single genome of linked genes already exists. Clearly, after one or a few replicating molecules emerged and linked, genomes began to grow in length at different rates in different lineages, up to consisting of billions of nucleotide base pairs in eukaryotes. The finding that the typical response to change in the fitness landscape is one of genome growth suggests an underlying mechanism for this rise in complexity. For example, Fig. 2.13 shows the simple case of $N = 1$ and $K = 0$ that experiences a change in its whole fitness landscape every 1000 generations, with $G = 1$. A constant, steady increase in genome

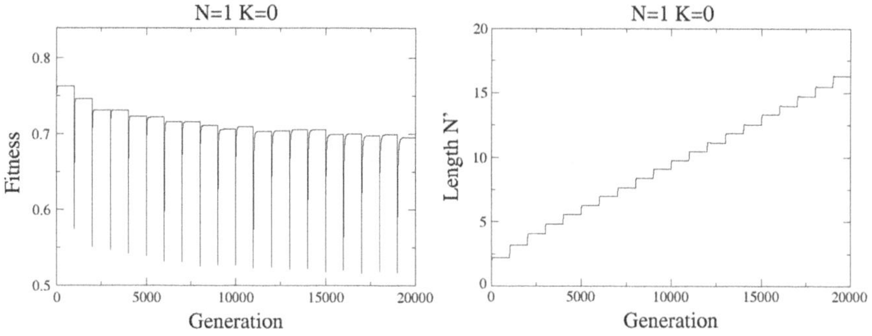

Fig. 2.13 Showing the typical behaviour over time when the whole fitness landscape is randomly recreated every 1000 generations starting from an initial length of $N = 1$, with $K = 0$ and $G = 1$

length is seen. Note the slight drop in fitness from those of the very low values of N is due to variance in the relatively small sample of random numbers from the expected optimum of 0.66 mentioned above.

2.3 Discussion

This chapter has considered the effects of fitness landscape ruggedness on the evolution of genome length in haploid single-celled organisms under different conditions. Using a simple, abstract model it has been shown that increases in genome length due to the addition of randomly created novel sequences which have an immediate effect upon fitness is the norm. That is, fitness landscape ruggedness does not hinder genome growth and can actually promote it as the window of opportunity for a randomly created sequence to have a beneficial—or at least neutral—effect on fitness increases due to the typical (low) height of peaks in such landscapes. It can also enhance fitness since it enables the avoidance of the complexity catastrophe for higher levels of gene epistasis. Note no explicit functional benefit from increased genome length was included in the model and hence the growth seen was as a consequence of the underlying dynamics of evolution on fitness landscapes of varying degrees of ruggedness *only*.

References

1. Bull, L.: Coevolutionary species adaptation genetic algorithms: a continuing SAGA on coupled fitness landscapes. In: Capcarrere, M., et al. (eds.) Proceedings of the Eighth European Conference on Artificial Life, pp. 322–331. Springer, Berlin (2005)
2. Crombach, A., Hogeweg, P.: Evolution of evolvability in gene regulatory networks. PLoS Comput. Biol. **4**(7), e1000112 (2008)

3. Deline, B., Greenwood, J., Clark, J., Puttick, M., Peterson, K., Donoghue, P.: Evolution of metazoan morphological disparity. Proc. Natl. Acad. Sci. USA **115**(38), E8909–E8918 (2018)
4. Harvey, I.: Species adaptation genetic algorithms: a basis for a continuing SAGA. In: Varela, F.J., Bourgine, P. (eds.) Toward a Practice of Autonomous Systems: Proceedings of the First European Conference on Artificial Life, pp. 346–354. MIT Press, Cambridge, MA (1992)
5. Hughes, A.: The evolution of functionally novel proteins after gene duplication. R. Soc. Lond. B **256**, 119–124 (1994)
6. Kauffman, S.A.: The Origins of Order: Self-organisation and Selection in Evolution. Oxford University Press, New York, NY (1993)
7. Kauffman, S.A., Levin, S.: Towards a general theory of adaptive walks on rugged landscapes. J. Theor. Biol. **128**, 11–45 (1987)
8. Maynard Smith, J., Szathmary, E.: The origin of chromosomes 1: selection for linkage. J. Theor. Biol. **164**, 437–466 (1993)
9. Ohno, S.: Evolution by Gene Duplication. Springer, New York (1970)
10. Otto, S., Whitton, J.: Polyploid incidence and evolution. Annu. Rev. Genet. **34**, 401–437 (2000)

Chapter 3
Symbiosis

Symbiosis represents evolution bringing together the genomes of different species and it is well-established that the phenomenon has been of great significance (e.g., see [11]). When the relationship between the symbionts evolves in the direction of increasing dependency, "a new formation at the level of the organism arises—a complex form having the attributes of an integrated morphophysiological entity" [9, p. 5]. This chapter explores the effects of fitness landscape ruggedness and connectedness upon the evolution of symbiotic organisms which live in close association.

Under the most intimate of symbiotic associations—endosymbiosis—one of the partners, the host, incorporates the other(s) internally. Endosymbionts can occur within or outside their host's cells. Extracellular endosymbionts can be either between the cells of host tissue, or in an internal cavity, such as the gut. Intracellular endosymbionts are usually enclosed by a host membrane. This chapter begins by exploring the conditions under which a general form of endosymbiotic relationship between two and then three coevolving species proves beneficial. This is done using Kauffman and Johnsen's [7] abstract NKCS model, which allows for the systematic alteration of various aspects of a coevolving environment, including landscape ruggedness, landscape connectedness, the degree of host control over the symbiont, and the relative rates of evolution of the partners. Increasing integration between the partners, symbiogenesis, also delineates the transfer of genes from one symbiont's genome to another—horizontal gene transfer—creating a more complex genome for the recipient. The evolutionary performance of endosymbionts who transfer increasing fractions of their genome to their partner is also explored. Symbiogenesis is seen to be beneficial across the parameter space of the model.

L. Bull, *The Evolution of Complexity*, Emergence, Complexity and Computation 37,
https://doi.org/10.1007/978-3-030-40730-8_3

3.1 The NKCS Model

Kauffman and Johnsen [7] extended the NK model (Sect. 2.1) and introduced the abstract NKCS model to enable the study of various aspects of coevolution. At an abstract level coevolution can be considered as the coupling together of the fitness landscapes of the interacting species. Hence the adaptive moves made by one species in its fitness landscape causes deformations in the fitness landscapes of its coupled partners. In their model, each gene is also said to depend upon C randomly chosen traits in each of the other S species with which it interacts. The adaptive moves by one species may deform the fitness landscape(s) of its partner(s). Altering C, with respect to N, changes how dramatically adaptive moves by each species deform the landscape(s) of its partner(s). Again, for each of the possible $K + (S \times C)$ interactions, a table of $2^{(K+(S \times C)+1)}$ fitnesses is created for each gene, with all entries in the range 0.0–1.0, such that there is one fitness for each combination of traits. Such tables are created for each species (Fig. 3.1, the reader is referred to [6] for full details).

Figure 3.2 shows example results for one of two coevolving species where the parameters of each are the same and hence behaviour is symmetrical. Again, as in Chap. 2 with the NK model, a species is said to be converged and hence represented by a single, asexual haploid genome. All results reported in this chapter are the average of 10 runs (random start points) on each of 10 NKCS functions, i.e., 100 runs, for 20,000 generations. Here $0 \leq K \leq 10$, $1 \leq C \leq 5$, for $N = 20$ and $N = 100$.

When $C = 1$, Fig. 3.2 shows examples of the general properties of adaptation on such fitness landscapes identified by Kauffman (e.g., [6]) in the NK model, i.e., when $C = 0$, still hold, including the height of peaks typically found begins to fall as K increases relative to N. Figure 3.2 further shows how increasing the degree of connectedness (C) between the two landscapes causes fitness levels to fall significantly (T-test, $p < 0.05$) when $C \geq K$ for $N = 20$. That is, as $K \rightarrow N$ a high number of peaks of similar height typically exist in each of the fitness landscapes and so the effects of switching between them under the influence of C is reduced since

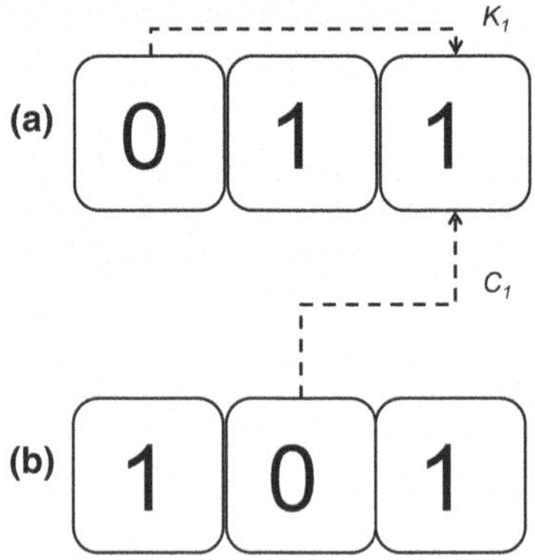

$K_0\ K_1\ C_1$	Fitness
0 0 0	0.76
0 1 0	0.23
1 0 0	**0.56**
1 1 0	0.98
0 0 1	0.43
0 1 1	0.32
1 0 1	0.65
1 1 1	0.03

Fig. 3.1 The NKCS model: Each gene is connected to K randomly chosen local genes and to C randomly chosen genes in each of the S other species. A random fitness is assigned to each possible set of combinations of genes. These are normalised by N to give the fitness of the genome. Connections and table shown for one gene in one species for clarity. $N = 1$, $K = 1$, $C = 1$, $S = 1$ here

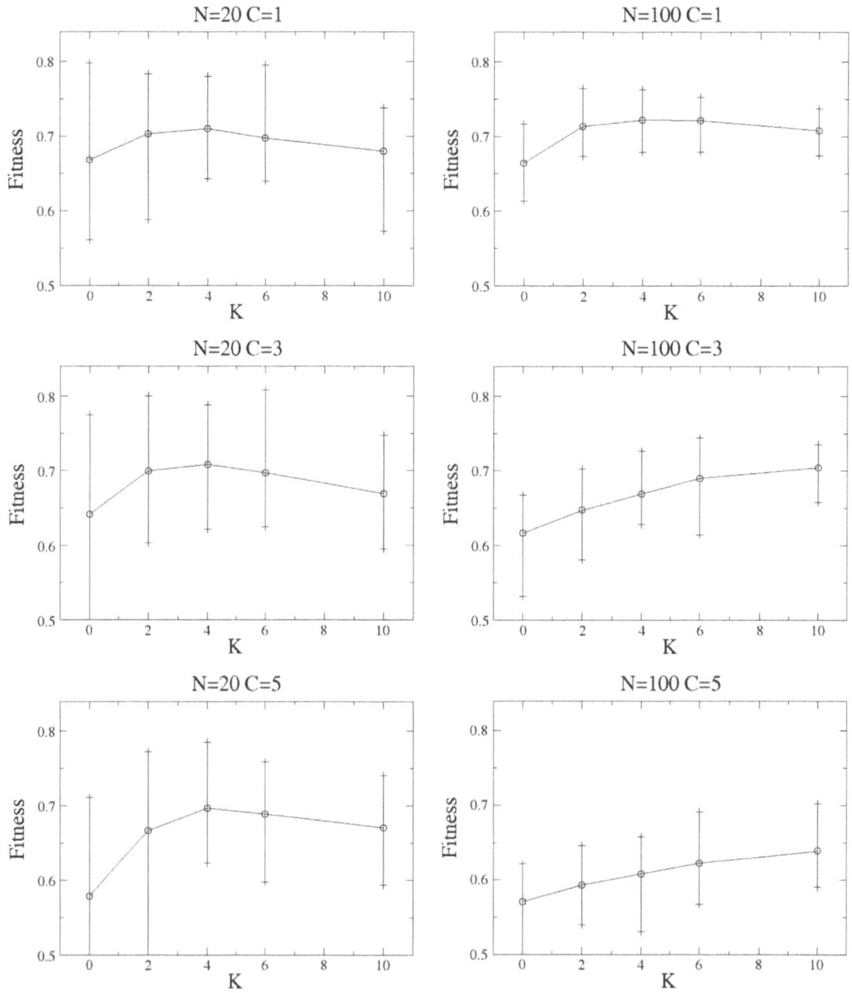

Fig. 3.2 Showing the fitness reached after 20,000 generations on landscapes of varying ruggedness (K), coupling (C), and length (N)

each landscape is very similar. Note this change in behaviour around $C = K$ was suggested as significant in [6] where $N = 24$ only was used throughout. However, Fig. 3.2 also shows how with $N = 100$ fitness *always* falls significantly with increasing C (T-test, $p < 0.05$), regardless of K. That is, it might be concluded that more complex organisms (>N) appear more sensitive to landscape coupling (>C).

As in [7], the models have also assumed that the two symbionts evolve at the same rate; each species coevolves in turn in the standard NKCS model. Following [1, 3], a new parameter R can be added to the model to represent the relative rate at which the endosymbiont evolves—by undertaking R rounds of mutation and selection to

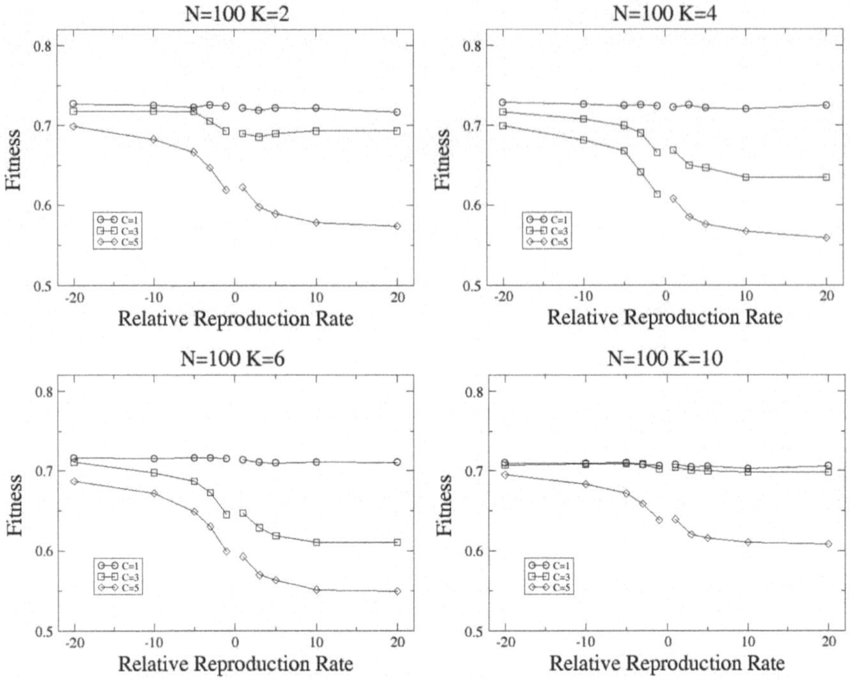

Fig. 3.3 Showing examples of the fitness reached after 20,000 generations for a species coevolving with a partner reproducing at a different relative rate R. Error bars are not shown for clarity

each round by the host. Figure 3.3 shows how with $N = 100$, generally, increasing R increases the effects of C for a given K when $C > 1$ whilst $R < 10$. In contrast, no notable effect is seen by increasing R for any K and C values tried when $N = 20$ (not shown). It can be noted that $N = 100$ (only) was also used in [1] and $N = 64$ (only) in [3].

3.2 Endosymbiosis in the NKCS Model

Endosymbiotic associations are often hereditary, wherein the host's endosymbiont(s) pass directly to its offspring. The mechanism for this perpetuation ranges from transmission in the egg cytoplasm to offspring ingesting the endosymbiont(s) shortly after birth, for example, when cows lick their calves thereby passing on their rumen ciliates. If symbiosis is viewed as organisms acquiring the properties of others, then hereditary endosymbiotic relationships in particular may be perceived as the direct inheritance of acquired characteristics, that is, as a Lamarckian-like process: "What is an [endosymbiont], from a genetic point of view? A new cluster of genes!" [13].

The above NKCS model can be modified slightly to capture an hereditary endosymbiotic relationship between the two coevolving partners. Simply, the second species can only accept a beneficial mutation if it does not decrease the fitness of the first (host) species, where ties are again broken at random. That is, the formation of an endosymbiosis is seen as an example of slavery with the endosymbiont being internalised purely for the benefit of the host (after [12]). This point is returned to later.

Figure 3.4 shows examples of the general result previously reported in [2] using a similar version of the NKCS model but extended to larger N: hereditary endosymbiosis proves beneficial to the host for all K and C (T-test, $p < 0.05$). That is, all inter-dependence (C) effectively becomes intra-dependence (K) and the host's fitness landscape no longer changes due to the effects of its symbiotic partner but instead becomes more rugged. This is particularly beneficial for low K partners coupled by high C due to the buckling-up effect from increasing K mentioned above.

There is some debate around whether (hereditary endosymbiotic) mitochondria facilitated or were acquired through phagocytosis in the evolution of eukaryotes due to its energetic cost (e.g., see [10]). If the capacity for phagocytosis is assumed to have increased relative genome size in the pre-eukaryotic cell, the results here suggest that the effects of its coevolutionary dependence upon the free-living mitochondria would have been more keenly felt: the fitness for a given K and C combination is typically significantly lower (T-test, $p < 0.05$) when $N = 100$ in comparison to $N = 20$ for $C > 1$ (Fig. 3.2), and the relative benefit of hereditary endosymbiosis is therefore larger for $N = 100$ (Fig. 3.4). The emergence of a complex trait such as phagocytosis may therefore have increased the chances of the uptake of mitochondria.

Having multiple, intimate symbiotic partners is not unusual: mealybugs contain cellular bacteria that contain their own bacteria, eukaryotes can contain chloroplasts alongside mitochondria, etc. Figure 3.5 shows example behaviour for the host with its hereditary endosymbiont described above, evolved in the context of another free-living symbiont (left) and the case where that symbiont has become a second hereditary endosymbiont (right), with $N = 20$. Following Burns' [4] suggestion that endosymbionts may typically no longer experience environmental factors directly, the effects of the free-living symbiont are not applied to the endosymbiont and vice versa here. As can be seen, in comparison to the equivalent case without the extra coevolving symbiont (Fig. 3.4), there is no significant change in the evolutionary behaviour for the host for all K and C combinations, even with a moderate amount of coupling ($W = 3$) to the extra symbiont. When $N = 100$, there is again a significant drop in fitness for the host when $K < 6$ and all C with $W = 3$ (T-test, $p < 0.05$, not shown). In all cases, the fitness of the host is increased by forming a secondary hereditary endosymbiosis with the symbiont (T-test, $p < 0.05$).

The reason for this change in behaviour with an increase in genome size is again as described above for Fig. 3.2 since, with a combined length of $2N$ here the host still exists on fitness landscapes where the degree of ruggedness—caused by *both* its own K and the C of its coupling to the endosymbiont—is high relative to N when $N = 20$ (combined $N = 40$). Thus its landscapes are very similar: the effects of moving from one to another due to evolutionary changes in its coupled partner are greatly

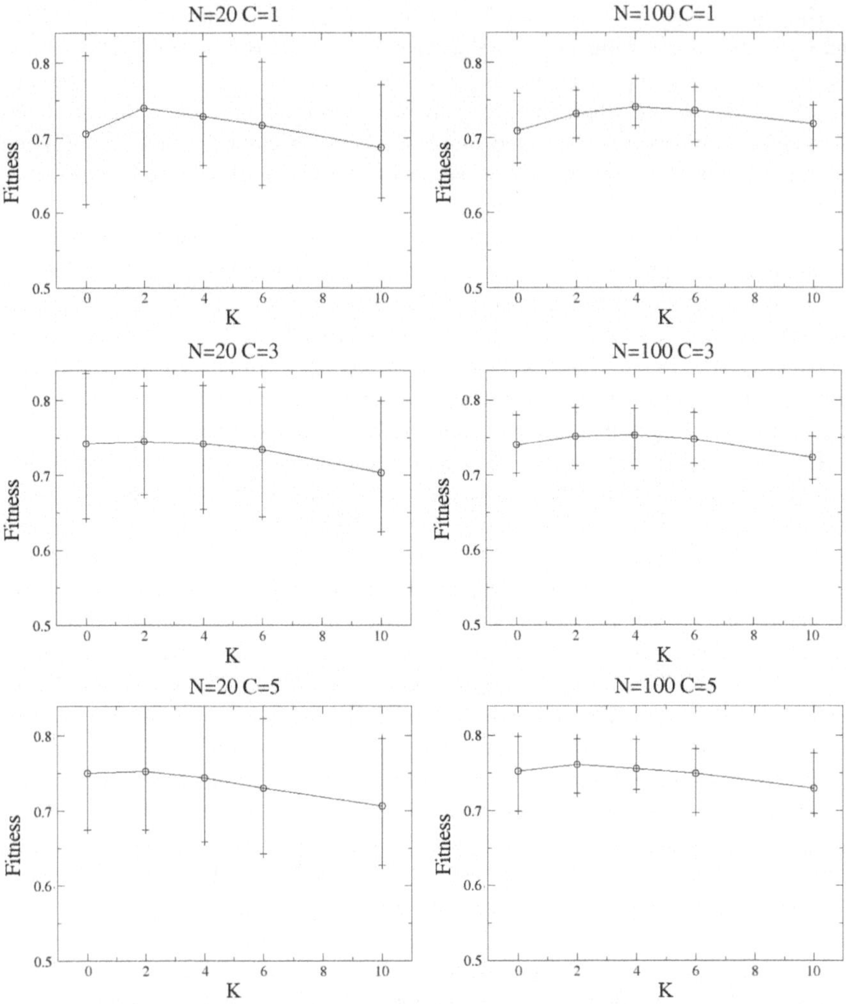

Fig. 3.4 Showing the fitness reached after 20,000 generations on landscapes of varying ruggedness (K), coupling (C), and length (N) when one of the two coevolving species is seen to permanently exist/evolve within the other, i.e., hereditary endosymbiosis

reduced as a similar height peak exists in close proximity in the new landscape as to the peak in the previous landscape. This suggests multiple intimate symbioses are more likely to emerge in host organisms evolving on more rugged fitness landscapes since they are less prone to suffering the effects of the prior coupling with their partner before enslaving them. However, organisms with longer genomes experience a greater improvement in fitness from enslaving the symbiont. Note how the process itself then places the host on a more rugged landscapes, etc.

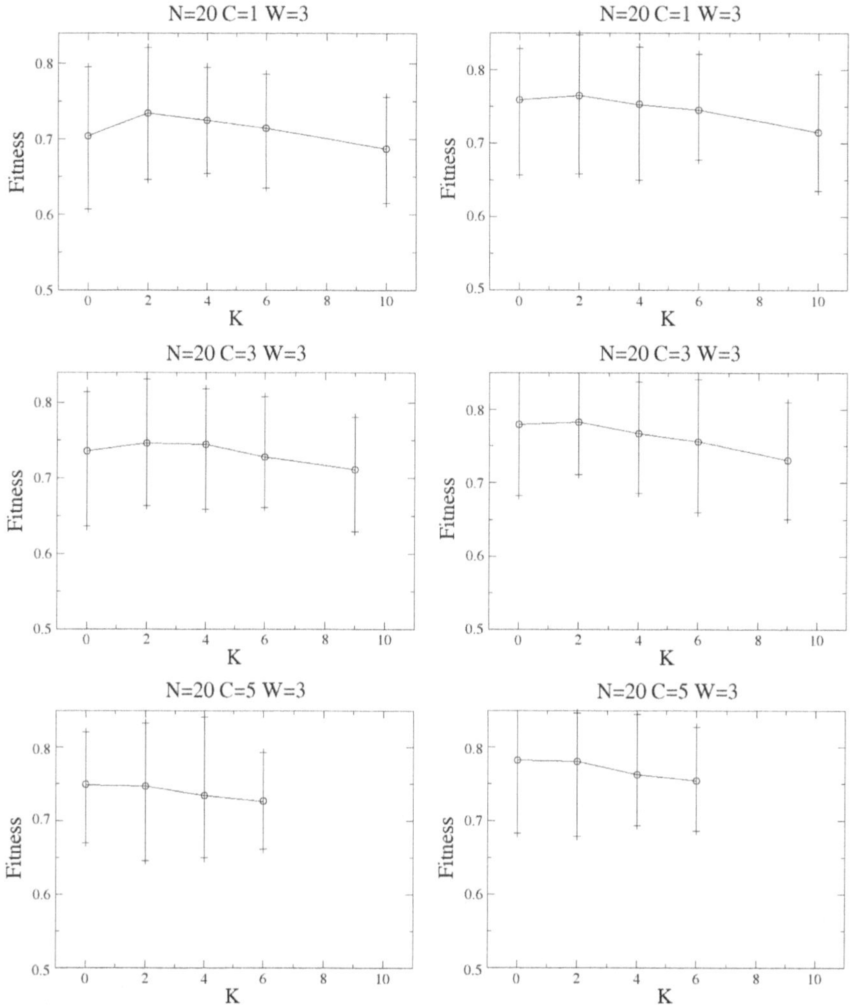

Fig. 3.5 Showing the fitness reached after 20,000 generations on landscapes of varying ruggedness (K) and connectedness (C) in an evolving environment to which it is connected ($W = 3$), for an hereditary endosymbiosis (left) and an hereditary endosymbiosis which further contains the environment/third species (right), i.e., plastid-like

The results are the same when the endosymbiont and free-living symbiont are coupled by W (not shown), as expected from Fig. 3.2.

As mentioned above, and as in [2], the assumption thus far has been that the endosymbiont's evolutionary progress is completely under the influence of the host in that no detrimental steps to its fitness are allowed. This is based on the idea that such relationships should be viewed as a form of enslavement by the host [12] rather than mutually beneficial (e.g., [11]). The consequences of reducing the host's

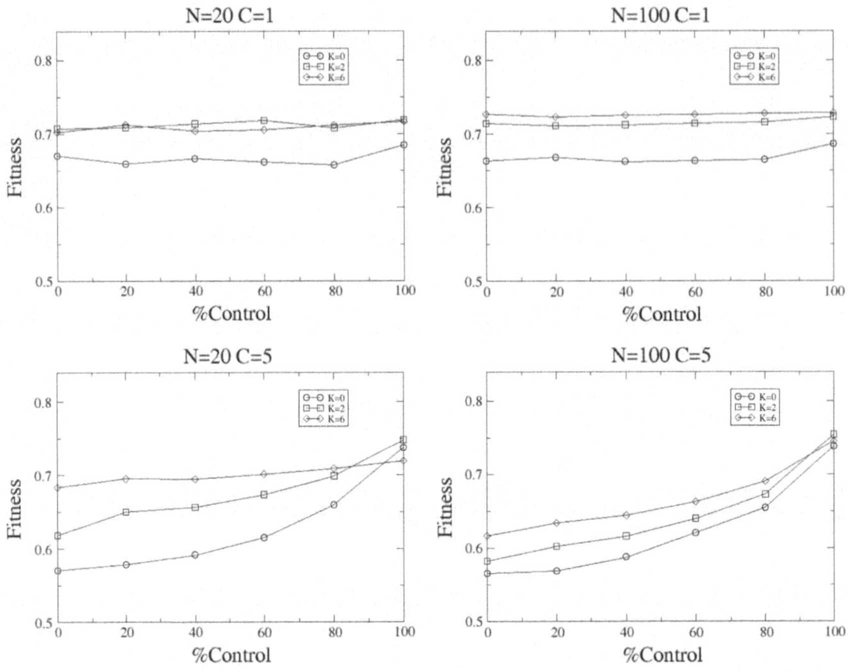

Fig. 3.6 Showing examples of the fitness reached after 20,000 generations when the host has different degrees of control over the evolutionary behaviour of its hereditary endosymbiont

level of control over the endosymbiont can be explored relatively easily within the model by introducing a probability for the acceptance of any host-detrimental but endosymbiont-beneficial mutations. That is, if such a case arises, the host-detrimental mutation is accepted for the endosymbiont if the given probability—the inverse of the host's percentage of control—is satisfied.

Figure 3.6 shows examples of how the host's fitness is affected depending upon its level of control. When it has no control (0%), fitnesses are as in the standard NKCS model shown above in Fig. 3.2, and when it has full control (100%), fitnesses are as above in Fig. 3.4. As can be seen, when $C = 1$, there is a slight benefit (T-test, $p < 0.05$) from increasing control from 80 to 100% when $K = 0$, regardless of N, otherwise there is no benefit in increased control over the endosymbiont. As noted above, since C effectively becomes K when the host has 100% control, the benefit seen is due to the aforementioned rise in optima height when $K > 0$. It is interesting to note that full control is required to gain that benefit, even though $C > K$. When C is further increased, the rate of control required to obtain a benefit to the host's fitness is reduced from 100%. For example, as shown in Fig. 3.6, regardless of N, when $C = 5$, a benefit is seen (T-test, $p < 0.05$) from over 20% control when $C > K$. When $N = 100$, a similar benefit is seen for all K. Again, the effects of $K \rightarrow N$ are seen for the smaller N.

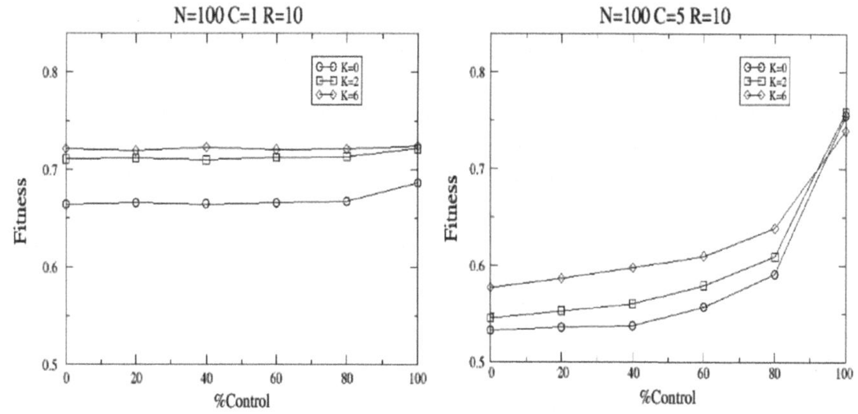

Fig. 3.7 Showing examples of the fitness reached after 20,000 generations when the host has different degrees of control over the evolutionary behaviour of its hereditary endosymbiont which is evolving ten times faster ($R = 10$)

As noted above, following [7] the models have assumed that the two symbionts evolve at the same rate; each species coevolves in turn in the standard NKCS model but a new parameter R can be introduced (Fig. 3.3). Figure 3.7 shows results from $R = 10$, i.e., the symbiont undertakes ten rounds of selection and mutation per one by the host. As can be seen, when $C = 1$ the results are the same as observed for $R = 1$ in Fig. 3.5. However, when $C = 5$, the host's fitness is significantly lower (T-test, $p < 0.05$) until its level of control is over 80% compared to the $R = 1$ case, and that no benefit from increased control is seen until it is over 40%. Hence a further, potentially significant benefit from forming an hereditary endosymbiosis is seen by the host if it also enables a reduction in the relative rate of evolution of the symbiont (see also [1]).

3.3 Horizontal Gene Transfer in Hereditary Endosymbiosis

As suggested above, symbiogenesis can be seen to act as a positive feedback loop in which each evolved beneficial adaption to the symbiosis further increases the association's chances of selection, thereby decreasing the partners' isolated fitness; the symbionts engaged in the association become increasingly obligate over evolutionary time (e.g., see [8]). Redundant traits are often selected against, which may include traits critical to the partners' isolated existence, further exaggerating this unifying process. A loss of genetic material is often seen within more intimate symbioses due to horizontal gene transfer: "[w]ith the transfer of genes, a symbiosis becomes more closely integrated. Part of the genome of one symbiont is transferred to the genome of the other. The new genome may underlie metabolic pathways leading to an advantageous product that neither partner was capable of producing alone" [11].

Within the NKCS model there is no scope for the emergence of novel functionalities but it can be used to examine the selective performance of gene transfer as a way of configuring interdependent genes. The original use of the model to explore this aspect of symbiogenesis assumed that *all* genes had been transferred to the genome of the host [2]. A later version of the model explored the effects of transferring a percentage T of the endosymbiont's genes to the host for various relative rates of evolution R between the two [1], where $0\% \leq T \leq 50\%$. Smaller values of T of up to 30% were found to be selectively neutral for the endosymbiont, becoming detrimental thereafter, regardless of R and for $C > 1$. However, it was assumed that the host had no selective control over the evolution of the endosymbiont, with it seeing a benefit from $T \geq 30\%$ for all R and $C > 1$.

Figure 3.8 shows examples of including horizontal gene transfer in the above model with varying levels of control over the endosymbiont's evolution by the host. The percentage T is taken from the left-hand end of the genome of the endosymbiont and placed onto the right-hand end of the host's genome. The fitness contribution of the transferred genes to the host are calculated using the same tables as before, i.e., from the endosymbiotic species' fitness function, and the total is now normalised by $N + T\%$, with the position of all the original K and C connections of the transferred genes maintained in both cases. Endosymbionts subsequently have their fitness contributions normalised by $N - T\%$.

As can be seen by comparing Figs. 3.8 to 3.6, no significant benefits are obtained by the host when C is low ($C = 1$) until $T = 90\%$ and then only in the $C > K$ case of $K = 0$, at a level of control of 80% or less (T-test, $p < 0.05$). When $C = 5$, $T = 30\%$ is beneficial when $C > K$ at a level of control of 80% or less. It is also beneficial for higher K with a level of control of 40% or less. As T is further increased, the host experiences increasing benefit for all levels of control less than 100%, such that by $T = 90\%$ there is no significant difference in fitness reached regardless of the level of control, as might be expected. That is, horizontal gene transfer can be seen as a mechanism through which the host is able to exert control over the endosymbiont. Other direct mechanisms of control might include physical space restrictions, expelling endosymbionts, etc. and the negative effect on selection for the host representing an indirect mechanism for control, of course.

The effects of increasing R have also been explored here. Figure 3.9 shows example results from $R = 10$, with the most marked change seen when $T = 30\%$ and $C = 5$ compared to the equivalent case when $R = 1$. Similarly to the results seen in Fig. 3.7, the increase in the relative rate of evolution in the endosymbiont causes a significant drop (T-test, $p < 0.05$) in fitness for the host until a significant level of control can be exerted, over 80% here for $K > 0$. The higher percentages of gene transfer appear able to negate the detrimental effects of less control thereafter.

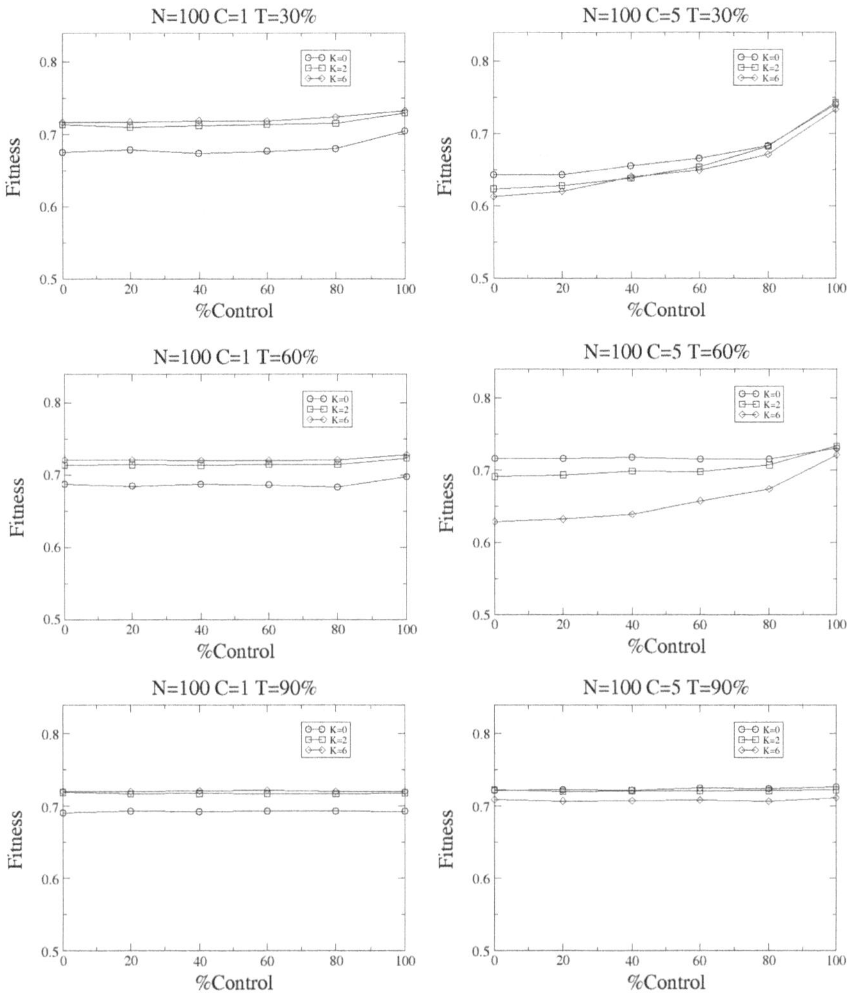

Fig. 3.8 Showing examples of the fitness reached after 20,000 generations when the host has different degrees of control over the evolutionary behaviour of its hereditary endosymbiont following different degrees of horizontal gene transfer (T) to the host

3.4 Discussion

Symbiosis is a fundamental process in nature and its role in the evolution of complexity is clear. Using the abstract NKCS model it has been shown that the formation of intimate symbioses is generally beneficial to the host since their inter-dependence becomes intra-dependence, i.e., the host's fitness landscape becomes more rugged and stable. In doing so, it has been shown that the genome length of the host can be significant with respect to the effects of the coupling between the symbionts before

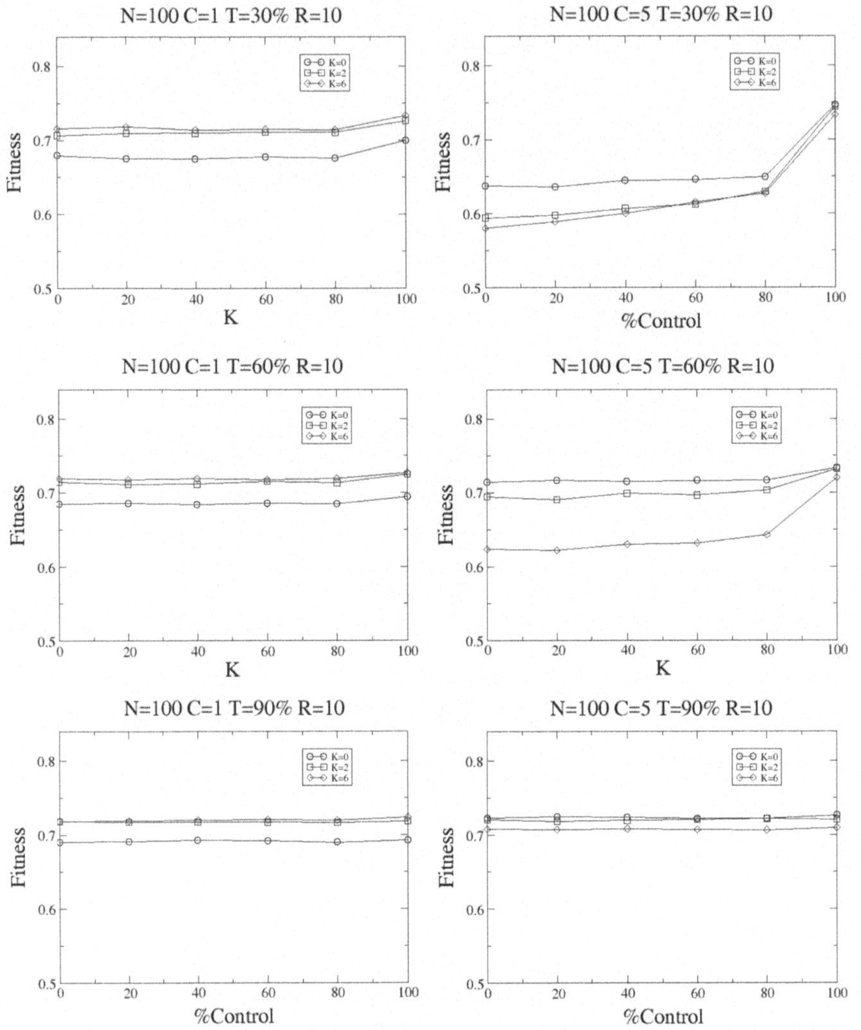

Fig. 3.9 Showing examples of the fitness reached after 20,000 generations when the host has different degrees of control over the evolutionary behaviour of its hereditary endosymbiont following different degrees of horizontal gene transfer to the host. The endosymbiont is evolving ten times faster ($R = 10$) than the host

they combine. In Chap. 2 the effects of fitness landscape ruggedness on the evolution of genome length in general were explored using the NK model, which included the findings that large (random) growth events can be expected to occur less often and that smaller growth events are more likely with increased ruggedness. Symbiogenesis represents a mechanism through which genome lengths can increase significantly in a single event through the inheritance of a pre-adapted set of genes and subsequently cause an increase in the ruggedness of the fitness landscape of the host. Such hosts have also been shown more likely to undergo a further symbiotic event. Thus symbiogenesis can indeed be seen to facilitate further increases in complexity.

It has been shown how the level of control of the host over the (enslaved) endosymbiont can be critical and horizontal gene transfer represents one mechanism by which that can be achieved. Mitochondrial DNA can contain anywhere from 3 to 67 genes and chloroplasts 60 to 100, with the assumption the rest of their original DNA has transferred to the nucleus, although some may simply have been lost. However, it is not clear why such organelle DNA remains (e.g., see [5]). The results here predict remaining organelle DNA since the findings indicate no potential benefit to the host from over 90% of genes being transferred in all cases and no significant benefit from over 60% of genes being transferred for lower levels of fitness landscape ruggedness.

References

1. Bull, L.: Artificial symbiogenesis and differing reproduction rates. Artif. Life **16**(1), 65–72 (2010)
2. Bull, L., Fogarty, T.C.: Artificial symbiogenesis. Artif. Life **2**(3), 269–292 (1996)
3. Bull, L., Holland, O., Blackmore, S.: On meme-gene coevolution. Artif. Life **6**(3), 227–235 (2000)
4. Burns, T.P.: Discussion: mutualism as pattern and process in ecosystem organisation. In: Kawanabe, H., Cohen, J.E., Iwaski, K. (eds.) Mutualism and Community Organisation, pp. 239–251. Oxford University Press, Oxford (1993)
5. De Grey, A.: Forces maintaining organellar genomes: is any as strong as genetic code disparity or hydrophobicity? BioEssays **27**, 436–446 (2005)
6. Kauffman, S.A.: The Origins of Order: Self-organisation and Selection in Evolution. Oxford University Press, New York, NY (1993)
7. Kauffman, S.A., Johnsen, S.: Co-evolution to the edge of chaos: coupled fitness landscapes, poised states and co-evolutionary avalanches. In: Langton, C.G., Taylor, C., Farmer, J.D., Rasmussen, S. (eds.) Artificial Life II, pp. 325–370. Addison-Wesley, Redwood City, CA (1992)
8. Keeler, K.H.: Cost: benefit models of mutualism. In: Boucher, D.H. (ed.) The Biology of Mutualism: Ecology and Evolution, pp. 100–127. Croom-Helm, London (1985)
9. Khakhina, L.N.: Concepts of Symbiogenesis: History of Symbiogenesis as an Evolutionary Mechanism. Yale University Press, New Haven, CT (1992)
10. Lane, N., Martin, W.: The energetics of genome complexity. Nature **467**(7318), 929–934 (2010)
11. Margulis, L.: Symbiosis in Cell Evolution. W.H. Freeman, Oxford (1992)
12. Maynard Smith, J., Szathmary, E.: The Major Transitions in Evolution. WH Freeman, Oxford (1995)
13. Nardon, P., Grenier, M.: Serial endosymbiosis theory and weevil evolution: the role of symbiosis. In: Margulis, L., Fester, R. (eds.) Symbiosis as a Source of Evolutionary Innovation, pp. 155–167. MIT Press, Cambridge, MA (1991)

Chapter 4
Sex

Whilst a number of explanations for various aspects of the evolution and maintenance of eukaryotic sex have been presented, none gives a unifying view of the wide variations in the process seen in nature. Sex is here defined as successive rounds of syngamy and meiosis in a haploid-diploid lifecycle. This chapter suggests that the emergence of a haploid-diploid cycle enabled the exploitation of a rudimentary form of the Baldwin effect (e.g., see [20] for an overview) and that this provides an underpinning explanation for all the observed forms of sex [5].

As discussed in [16, p. 150] the first step in the evolution of eukaryotic sex was the emergence of a haploid-diploid cycle, probably via endomitosis, before simple syngamy. Cleveland [8] was first to suggest that organisms may become diploid by a variation in mitosis to maintain the genome copy, i.e., endomitosis. Syngamy, the fusion of two independent genomes, probably emerged thereafter. The subsequent emergence of isogamy, i.e., mating types, is not considered in this chapter. Under both scenarios, a previously haploid cell became diploid. A number of explanations have been presented for why a diploid, or increasing ploidy in general, is beneficial, typically based around the potential for "hiding" mutations within extra copies of the genome (e.g., see [17] for an overview). A change in ploidy can potentially alter gene expression, and hence the phenotype, even if no mutations occur between the lower and higher ploidy states—through epigenetic mechanisms, through rates of changes in gene product concentrations, no or partial or co-dominance, etc. (e.g., see [7]). In all cases, whether the diploid is formed via endomitosis or syngamy, the fitness of the cell/organism is a combination of the fitness contributions of the composite haploid genomes. If the cell subsequently remains diploid and reproduces asexually, there is no scope for a rudimentary Baldwin effect. However, if there is a reversion to haploid cells under meiosis, there is potential for a mismatch between the utility of the haploids compared to that of the polyploid; individual haploids do not contain all of the genetic material over which selection operated. That is, the effects of genome combination can be seen as a simple form of phenotypic plasticity for the individual haploid genomes before they revert to a solitary state and hence the Baldwin effect may occur.

L. Bull, *The Evolution of Complexity*, Emergence, Complexity and Computation 37,
https://doi.org/10.1007/978-3-030-40730-8_4

4.1 The Baldwin Effect in the NK Model

Recall that in the NK model used throughout this book, the population is of size one and a species evolves by making a random change to one randomly chosen gene per generation. Following [3], a very simple (random) learning process to enable phenotypic plasticity can be added to evolution by allowing a new individual to make a further L (unique) mutations after the first. If the averaged fitness of this "learned" configuration and that of the first mutant is greater than that of the original, the species is said to move to the *first* mutant configuration but assigned the *averaged fitness* of the two configurations. All results reported are the average of 10 runs (random start points) on each of 10 NK functions, that is 100 runs, for 50,000 generations. Here $0 \leq K \leq 15$ and $0 < L \leq 7$, for $N = 20$ and $N = 100$.

Figure 4.1 shows the performance of the Baldwin effect across a wide range of K and L combinations for $N = 20$. For $K = 0$, the unimodal case, learning shows no benefit for evolution (T-test, $p \geq 0.05$, $0 < L < 7$) and is disruptive when applied at higher levels (T-test, $p < 0.05$, $L = 7$). As K increases, i.e., as landscape ruggedness increases, learning becomes beneficial across a wider range of L. When $0 < K < 6$, learning is either beneficial (T-test, $p < 0.05$, $0 < L < 7$), or has no effect (T-test, $p \geq 0.05$, $L = 7$). Learning is always beneficial over the ranges used when $K \geq 6$ (T-test, $p < 0.05$). The smallest amount of learning $L = 1$ is as beneficial as any other until $K > 6$, when the higher levels are most beneficial (T-test, $p < 0.05$, $L \geq 5$). The same results are typically seen for $N = 100$, although the higher amounts of learning are not found beneficial for $K > 4$ (Fig. 4.2). These findings support those reported in [3]: the most beneficial amount of learning varies with K.

As noted above, as well as the most useful amount of such random learning varying with the ruggedness of the fitness landscape, i.e., L, it was also shown how the frequency at which the simple learning process is applied to an individual can alter the conditions under which the Baldwin effect is beneficial [3]. Figure 4.3 shows examples from varying the frequency of learning both within and across lifecycles. In the previous results, the fitness of a genome was calculated as the average of its purely genetic configuration and that of the learned configuration. Thus learning can be seen to have occurred throughout the lifecycle. This can be varied such that the fitness of the learned configuration is weighted less equally to the genetic configuration: less learning. Examples of the case of learning being weighted at 50% are shown in the top row of Fig. 4.3. Results show learning is now beneficial for all L for $K = 2$, with no significant change in behaviour to Fig. 4.1 for $2 < K \leq 6$, whilst learning is no longer beneficial for $K > 6$ (T-test, $p \geq 0.05$). Figure 4.2 also shows examples from only allowing the original whole lifecycle learning to occur on every other generation. The results are the same as for the half lifecycle case, except there is no drop in benefit for $K > 6$.

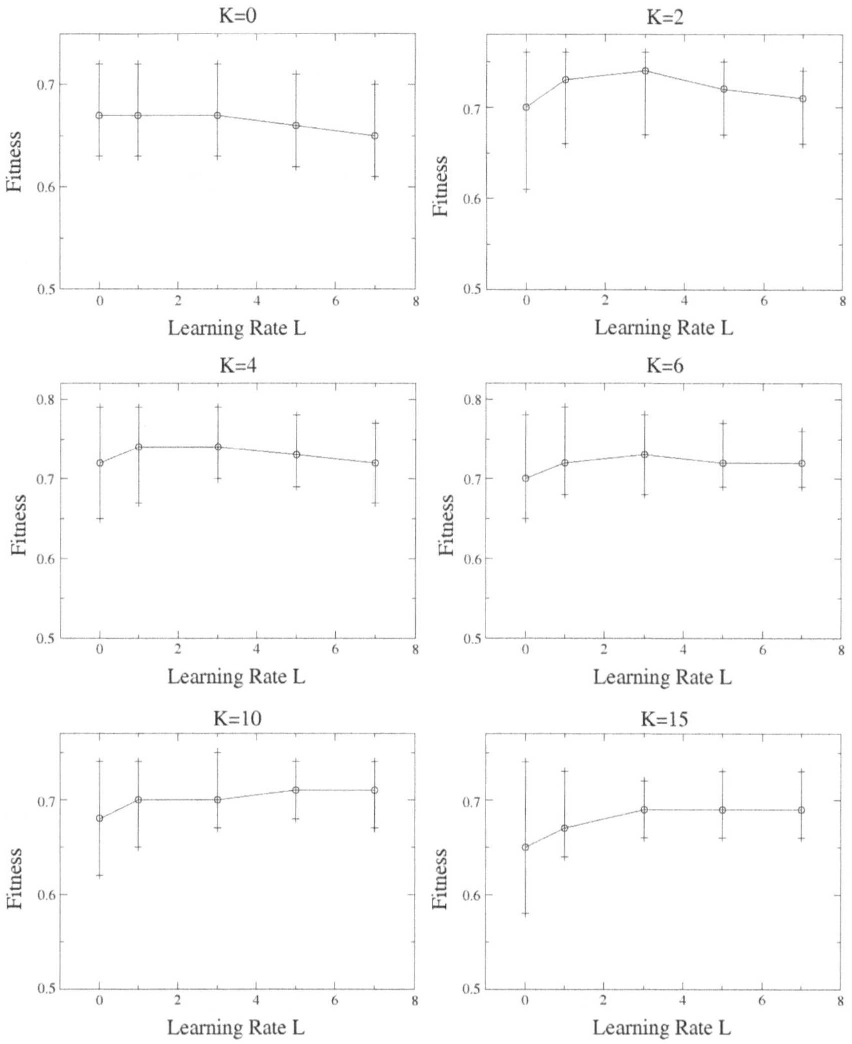

Fig. 4.1 Performance of the Baldwin effect, after 50,000 generations, for varying amounts of learning (L), on landscapes of varying ruggedness (K) with $N = 20$

4.2 Evolution of the Haploid-Diploid Cycle: The Baldwin Effect

Whether the haploid-diploid cycle emerged via endomitosis or via a simple form of syngamy is not crucial to the basic hypothesis presented here. As noted above, explanations primarily based around mutation hiding have been given as to why a diploid state is beneficial to a haploid state. Similarly, there are explanations for the

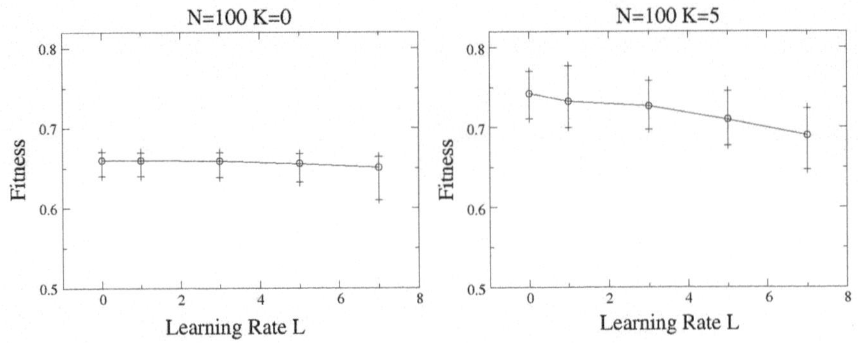

Fig. 4.2 Example performance for varying amounts of learning (L), on landscapes of varying ruggedness (K), with $N = 100$

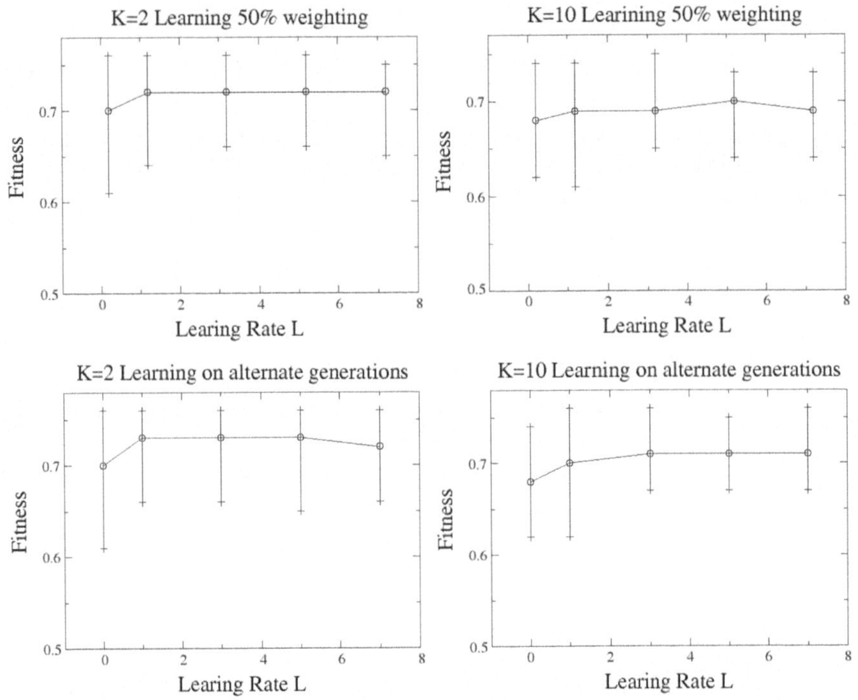

Fig. 4.3 Performance of the Baldwin effect, after 50,000 generations, where learning occurs at different frequencies ($N = 20$). The top row shows examples of learning occurring for half of the lifetime and the bottom row shows learning occurring every other generation

emergence of the alternation between the two states, typically based upon its being driven by changes in the environment (after [15]). If, as suggested here, the diploid state should be seen as the "learning" part of the lifecycle due to genome interactions, the results above anticipate the wide range of different haploid-diploid frequencies seen in nature. For example, most mammals have a primarily diploid lifecycle, many land plants exploit a (significant) haploid seed phase, etc. That is, as K and L vary, the optimal frequency with which learning occurs varies. Following [16, p. 150], endomitosis is assumed to have occurred first in this chapter.

The standard NK model (Sect. 2.1) is altered such that once the haploid genetic mutant is created, a copy is made and another gene chosen for further mutation (Fig. 4.4, left). Both genomes are evaluated and, for simplicity, the fitness of the diploid is assigned as their average. If the diploid is fitter than the diploid representing the current population, the species is said to move to the new configuration. Again for simplicity, selection picks one of the two genomes of the diploid at random. For higher levels of ploidy (four and eight) explored, the copy and mutation process is repeated equally for each new genome to the required level. That is, the rounds of endomitosis can be seen as rounds of learning by the cell/organism.

Figure 4.5 shows examples of how a haploid-diploid cycle via endomitosis is beneficial over a purely haploid (non-learning) cycle for all $K > 0$ (T-test, $p < 0.05$). It can also be seen that a further round of endomitosis to a tetraploid state before meiosis provides no benefit over diploidy for any K (T-test, $p \geq 0.05$), except when $K = 4$ (T-test, $p < 0.05$), with another round to octaploidy providing no benefit for low K and becoming detrimental for $K \geq 6$ (T-test, $p \geq 0.05$). The same behaviour was found for $N = 100$ (not shown).

The case for a haploid-diploid being beneficial was predicted above since the endomitotically produced genome is the same as the $L = 1$ case which was found to be beneficial for all $K > 0$. That the tetraploidy case is beneficial over the haploid for higher K is also anticipated by the previous results. However, whilst tetraploidy and octaploidy may be seen as the $L = 3$ and $L = 7$ cases respectively, they are subtly different. In the basic model, all L random learning changes are made in *one* copy of the genome. In the polyploidy cases, each further random learning change is made in genomes copied from genomes which have already had changes made. Hence increasing ploidy both increases the distance learning can sample from the original evolution produced genome point in the fitness landscape *and* the number of learning samples. For example, two genomes have $L = 1$ and one $L = 2$ in the tetraploid case. Thus increasing the number of samples also appears disruptive, even for lower range L.

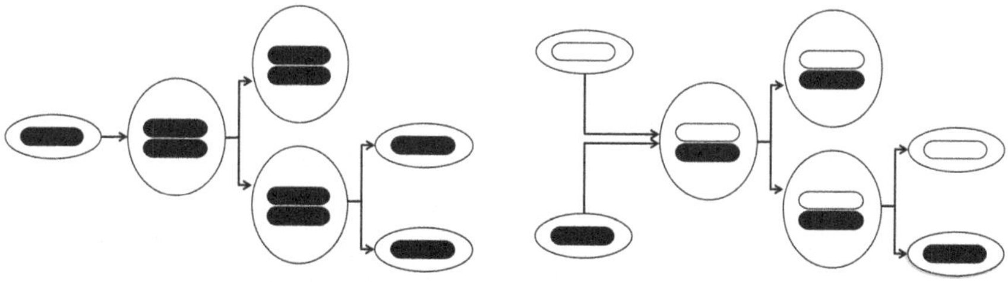

Fig. 4.4 The endomitosis (left) and syngamy (right) processes explored here (after [16, p. 151])

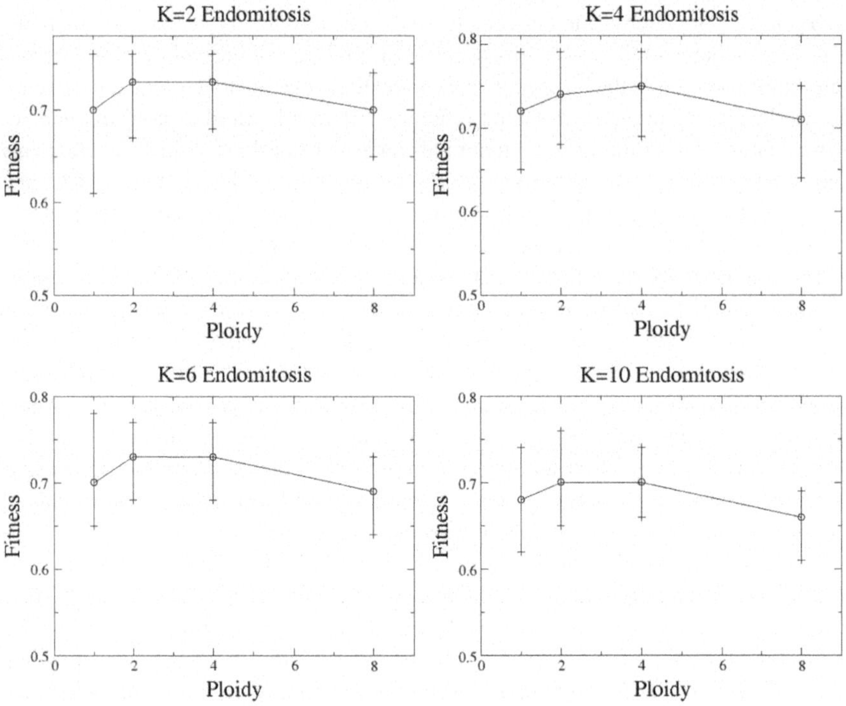

Fig. 4.5 Performance of the Baldwin effect under endomitosis, after 50,000 generations, for varying amounts of ploidy/learning, on landscapes of varying ruggedness (K) with $N = 20$

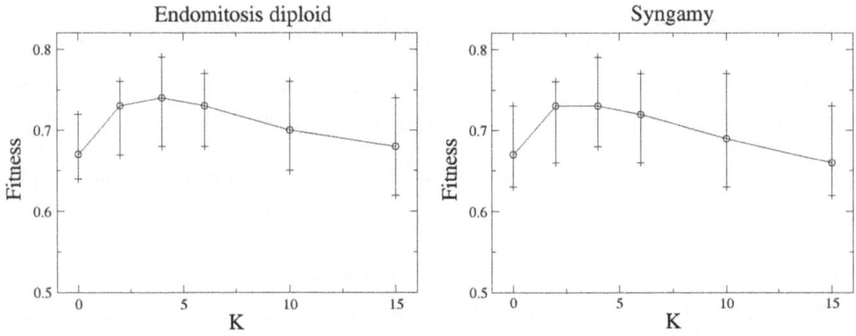

Fig. 4.6 Comparing the performance of the Baldwin effect under endomitosis and syngamy, after 50,000 generations, on landscapes of varying ruggedness (K) with $N = 20$

Figure 4.6 shows the comparative performance of the haploid-diploid cycle under endomitosis from Fig. 4.5 to that of the equivalent simple syngamy case. In the latter, the new diploid is created either by copying and mutating one gene in each of the species' two genomes, or by copying either genome twice and then mutating each once (Fig. 4.4). Both genomes are initialized as the same.

Figure 4.6 shows how there is no difference between either mechanism to provide the diploid stage for $K < 6$, whereafter endomitosis proves more beneficial (T-test, $p < 0.05$). The reason for this difference is again due to the difference in the amount of learning occurring per cycle; the results in Fig. 4.1 indicate a general benefit from an increased amount of learning with increasing fitness landscape ruggedness. In the endomitosis case, the learning change is added onto the genetic mutation of the first offspring genome in the second offspring genome. In the syngamy case, both genomes undergo the first genetic mutation change only. When the same genome is chosen twice to form the diploid, the syngamy case's sampling distance in the fitness landscape from the evolutionary origin is reduced in comparison to the equivalent endomitosis case (by one mutant step). When the two genomes are different in the syngamy case, this is not necessarily true, depending upon the degree of genetic diversity between the two original haploid genomes. As above, when $N = 100$, the extra learning—of endomitosis—provides no extra benefit and both perform equally well for all K (not shown). Comparison with an equivalent asexual diploid finds both endomitosis and syngamy more beneficial for all $K > 2$ (T-test, $p < 0.05$, not shown). This is also true even if all three possible diploid combinations from the two haploid genomes are evaluated per generation (T-test, $p < 0.05$, not shown). As noted above, no Baldwin effect can occur.

The type of Baldwin effect working here can be seen to alter the general characteristics of the evolutionary process. In the traditional haploid view of evolution, variation operators such as mutation copy errors, gene transfers, etc., generate a new genome at a *point* in the fitness landscape for evaluation. Under the haploid-diploid cycle, the variation operators create the bounds for sampling a *region* within the haploid fitness landscape by specifying two end points, i.e., each haploid genome to be partnered in the diploid. The actual position of the fitness point for the (diploid) phenotype taken from within that region then depends upon the percentage of the lifecycle the diploid state occupies—the larger, the closer to the midpoint (with all other things being equal) in the haploid landscape. Significantly, evolution assigns a single fitness value to the region of the fitness landscape the two haploid genomes delineate—evolution can be seen to be *generalizing* over the space.

This explains the increased benefit of the haploid-diploid cycle seen above as landscape ruggedness is increased. It can also be noted that the shape of the fitness landscape *varies* based upon the haploid genomes which exist within a given population at any time and how they are paired: sexual selection can be seen as a mechanism through which this is exploited. This is also significant since, as has been pointed out for coevolutionary fitness landscapes [4], such movement potentially enables the *temporary* creation of neutral paths, where the benefits of (static) landscape neutrality are well-established [12].

4.3 Two-Step Meiosis and Recombination: Altering the Amount of Learning

The few explanations as to why a form of meiosis exists which includes a genome doubling stage—the diploid temporarily becomes a tetraploid—range from DNA repair (e.g., [1]) to the suppression of potentially selfish/damaging alleles (after [10]). Explanations for the recombination stage vary from the removal of deleterious mutations (e.g., [13] to avoiding parasites (after [11]) (see [2] for an overview). With the Baldwin effect view proposed here, such sexual reproduction can be seen as a mechanism through which to vary the amount of learning a cell/organism can exploit during the diploid phase. The role of recombination becomes clear under the Baldwin effect view: recombination moves the current end points in the underlying haploid fitness space which define the generalization either closer together or further apart. That is, recombination adjusts the size of an area assigned a single fitness value, potentially enabling higher fitness regions to be more accurately identified over time. Moreover, recombination can also be seen to facilitate genetic assimilation within the simple form of the Baldwin effect. That is, the pairing of haploid genomes is seen as a "learning" step with the fitness of a given haploid affected by the allele values of its partner. If the pairing is beneficial and the diploid cell/organism is chosen under selection to reproduce, the recombination process brings an assortment of those partnered genes together into new haploid genomes. In this way the fitter alleles from the pair of partnered haploids may come to exist within individual haploids more quickly than under mutation alone.

The previous model of syngamy with one-step meiosis can been extended such that the two parental haploid genomes each become a gamete alongside their one-mutant genomes, which are also recombined (randomly chosen single point crossover) with each other. Two of the four resulting haploid genomes/gametes are then chosen at random to create the cell/organism for fitness evaluation (Fig. 4.7, left). Both haploids are again initialised as the same here to consider the emergence of the

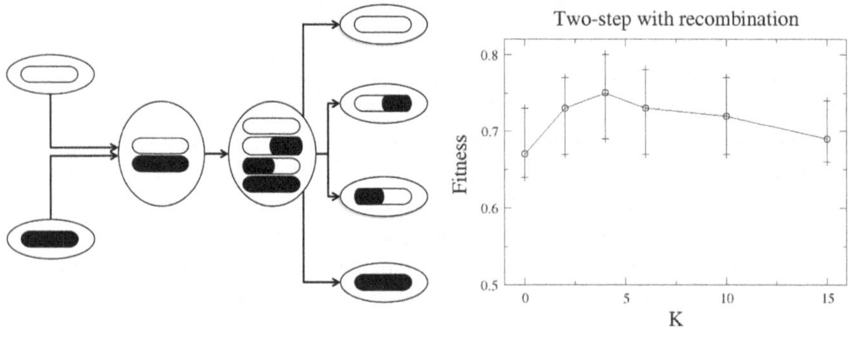

Fig. 4.7 Two-step meiosis with recombination process (left) and its performance, after 50,000 generations, on landscapes of varying ruggedness (K) with $N = 20$ (right)

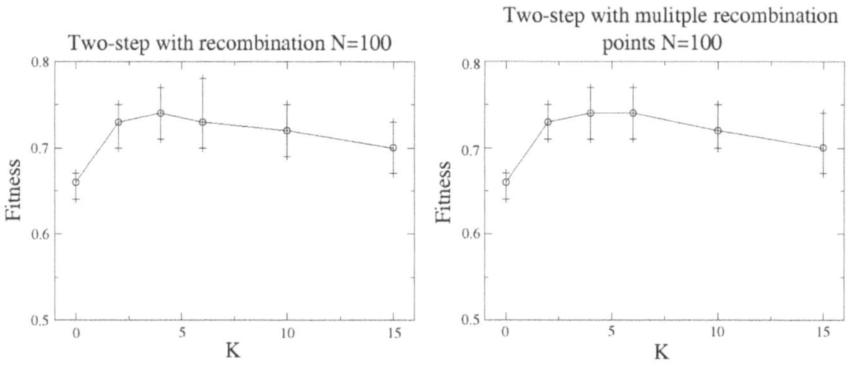

Fig. 4.8 Showing the comparative performance of two-step meiosis with a single-point (left) and multi-point (right) recombination processes, after 50,000 generations, on landscapes of varying ruggedness (K) with $N = 100$

process, a limitation removed in subsequent sections/chapters. Figure 4.7 (right) shows the typical behaviour for various K. In comparison with both endomitosis and syngamy, it is found that the increased learning is beneficial for all $K > 2$ (T-test, $p < 0.05$), as anticipated by the results in Sect. 4.1. The same general results as before are found for $N = 100$—the extra learning provides no benefit (not shown). However, greatly increasing the number of possible recombination points such that each gene is swapped with equal probability, means improved performance is again seen (Fig. 4.8) for $K > 4$ (T-test, $p < 0.05$, $K = 6$; $p \leq 0.10$, $K = 10, 15$). That is, the increased potential variation in the size of the generalization (end positions) is required for the larger fitness landscape. Similarly, a significant drop in performance is seen for all cases when recombination is removed (not shown).

As noted above, the percentage of their lifecycle eukaryotes spend as diploids varies greatly across species. Similarly, some species alternate between being sexual and asexual, such as aphids. Following the results in Fig. 4.3 (top), the case of half the lifecycle being spent as a haploid can be considered. Here one of the two haploids is chosen at random to make a 50% contribution to the fitness of the offspring, with the other 50% determined as the pair's average, as before. Figure 4.9 (left) shows how there is no significant difference in fitness for $K < 10$ (T-test, $p \geq 0.05$) but a significant decrease in fitness is seen for $K \geq 10$ (T-test, $p < 0.05$) compared to the animal-like diploid lifecycle case considered in Fig. 4.7. Other percentages of time as a haploid have not been explored here but the results in Fig. 4.3 suggest beneficial weightings exist. Figure 4.2 (bottom) also showed the potential benefits of varying the frequency of learning. Figure 4.9 (right) shows how varying the frequency of sexual reproduction to asexual reproduction, where the diploid is mutated once in each haploid to form an offspring in the latter, provides an increase in fitness at a ratio of asexual generations to one sexual generation of 7:1 (T-test, $p \leq 0.10$) for $K = 2$, with no significant change otherwise. No benefit was found for the other ratios and values of K explored (not shown). However, it can also be noted that if

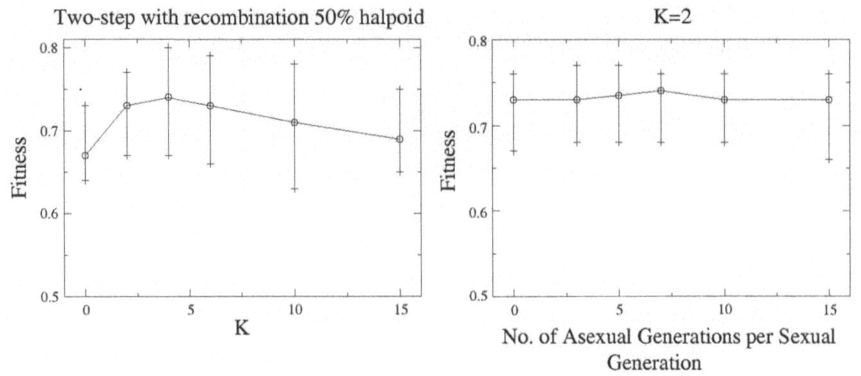

Fig. 4.9 Showing two-step meiosis with recombination process where the lifecycle is 50% haploid and its performance, after 50,000 generations, on landscapes of varying ruggedness (K) with $N = 20$ (left). And showing two-step meiosis with recombination with varying rounds of asexual reproduction per sexual reproduction event (right)

environmental conditions vary temporally such that the underlying ruggedness of the species' fitness landscape is increased/decreased, sexual reproduction is likely to be more/less effective at that time.

4.4 Genome Growth in Sexual Diploids

Chapter 2 explored the effects of extending the basic NK model such that genome size could vary under mutation for asexual diploids. Figure 4.10 shows the effects of enabling genome growth in sexual diploids, with $G = 1$. Note each haploid genome again either experiences a gene allele mutation or gene addition event, with equal probability. The recombination point is taken from within the range of the shorter of the two genomes when they are of different length. As can be seen, fitness levels are higher for all $K > 0$ in comparison to the equivalent haploid case (Fig. 2.3) (T-test, $p < 0.05$) and there is an increased amount of growth in all cases (T-test, $p < 0.05$). Figure 4.11 shows the underlying dynamics of the evolutionary process with the diploids. The general finding from Chap. 2 that genome length variation continues for longer with increasing landscape ruggedness is again seen here. However, evolution continues for much longer than for the haploids (Fig. 2.4). Figure 4.12 shows how this longer period of evolution creates the conditions for the adoption of many novel sequences, particularly early on (contrast with Fig. 2.5).

The diploids of course contain $2N$ genes. To explore the reason(s) for the increased levels of growth, asexual haploids of $2N$ genes have also been evolved with $G = 1$ but similar lengths were seen as with N genes (not shown). Figure 4.13 shows an example of evolving equivalent asexual diploids, i.e., diploids which do not undergo the haploid-diploid cycle with meiosis. Results indicate that similar lengths evolve

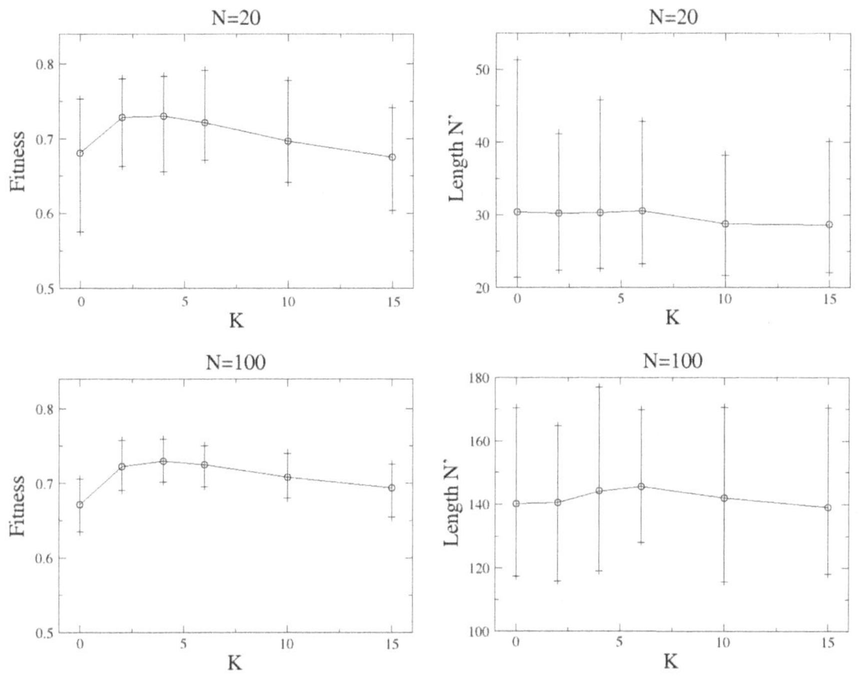

Fig. 4.10 Showing the fitness and length reached by diploids after 20,000 generations on landscapes of varying ruggedness (K) where the initial size (N) can increase by one gene under mutation

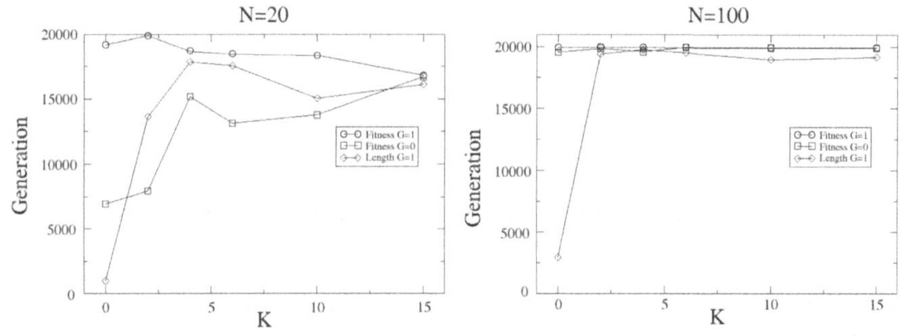

Fig. 4.11 Showing mean walk length to an optimum for a given N and K in sexual diploids, $G = 1$

as in the equivalent sexual diploid case and therefore neither the Baldwin effect mentioned nor recombination can be seen to be significantly contributing to the change in behaviour. As noted above, fitness levels are lower for $K > 2$ in the asexual diploid case (T-test, $p < 0.05$), even if all three one-mutation combinations for the two haploid genomes are tried per generation (not shown). Hence the main reason for the relative increase in genome length appears to be the change in dynamic(s)

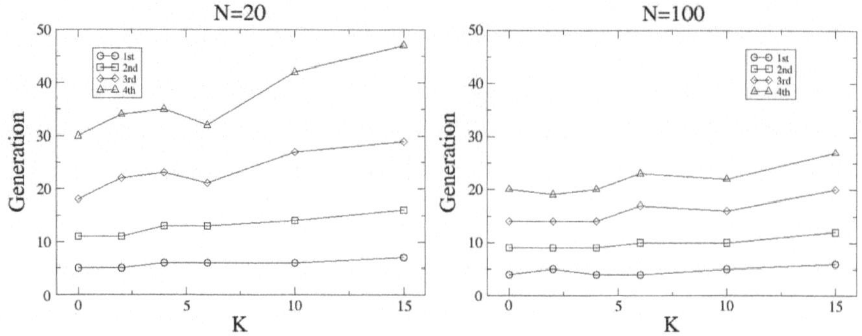

Fig. 4.12 Showing how the waiting time for each increase in length for a given N and K increases, $G = 1$

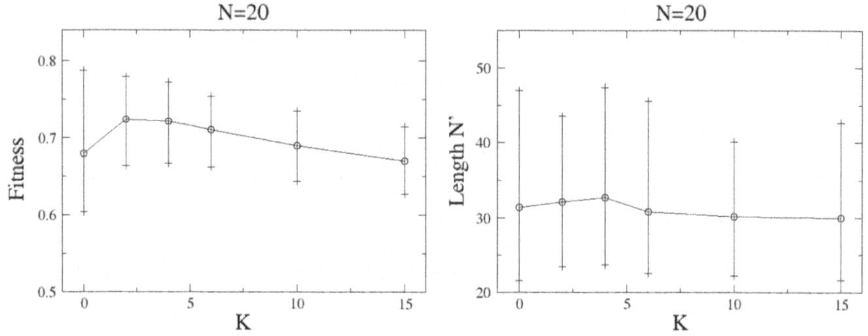

Fig. 4.13 Showing the size and fitness reached by asexual diploids after 20,000 generations on landscapes of varying ruggedness (K) where the initial size (N) can increase by one gene under mutation

experienced when two haploid genomes contribute to the fitness level. This is the case even if only one of the two genomes is allowed to experience an addition event per generation—although slightly reduced lengths are seen (not shown).

When $G = 20$, evolved genome lengths in the asexual haploids and sexual diploids are roughly equivalent (Fig. 2.6). That is, the period of time where changes in genome length are seen in diploids becomes similar to that seen in the haploids and hence a similar number of novel sequences are adopted in the time, particularly for $N = 20$. Thereafter, the general behaviour is the same for the diploids as the haploids (not shown) in that the most marked difference between $G = 1$ and $G = 20$ is again an increase in fitness for $K > 6$ (T-test, $p < 0.05$) when $N = 20$ (Fig. 4.14).

Similarly, when genome lengths can both increase and decrease by G, fitness levels remain unaffected but significantly less growth is seen for all K such that there is no difference between the sexual diploids and asexual haploids (not shown). Figure 4.15 shows how the sexual diploids are more sensitive to the presence of

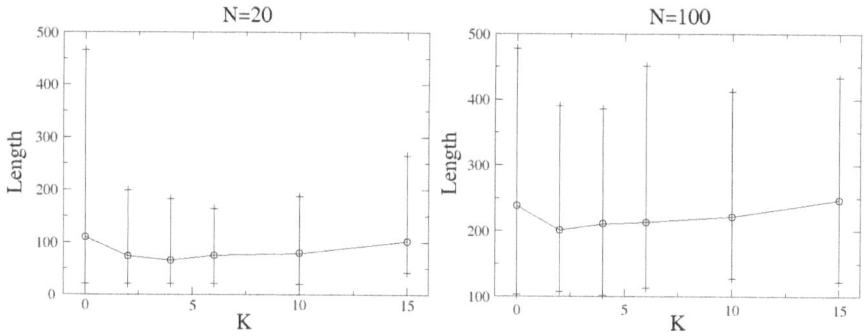

Fig. 4.14 Showing the length reached by sexual diploids after 20,000 generations on landscapes of varying ruggedness (K) where the initial size (N) can increase by twenty genes under mutation

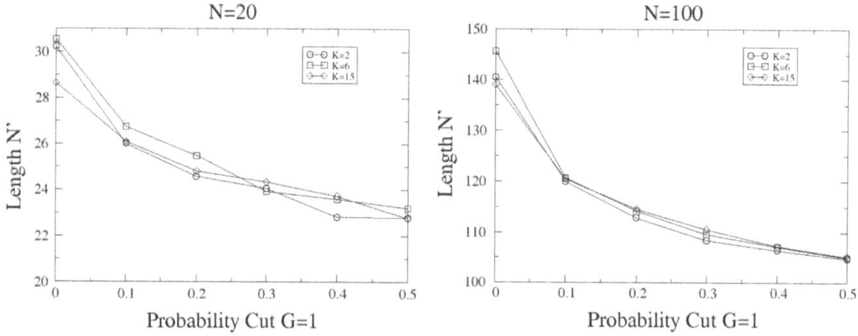

Fig. 4.15 Showing the effect of varying the probability of decreasing the number of genes for different N and K in sexual diploids (contrast to Fig. 2.10)

a deletion process in comparison to the asexual haploids, with lengths decreasing rapidly as the probability increases.

The results and behaviour on the changing landscapes are generally the same as with the asexual haploids (not shown).

4.5 Coevolving Sexual Diploids

As in Chap. 3, all previous known studies of coevolution with the NKCS model have used asexual haploid species. Given the results in this chapter within the NK model, the behaviour of sexual diploid species in the NKCS model can be compared with the typical behaviour seen with asexual haploids and to determine where sex is beneficial or not within a coevolutionary context. Figure 4.16 shows examples of extending the NKCS model such that one of a pair of species is a sexual diploid with the other an

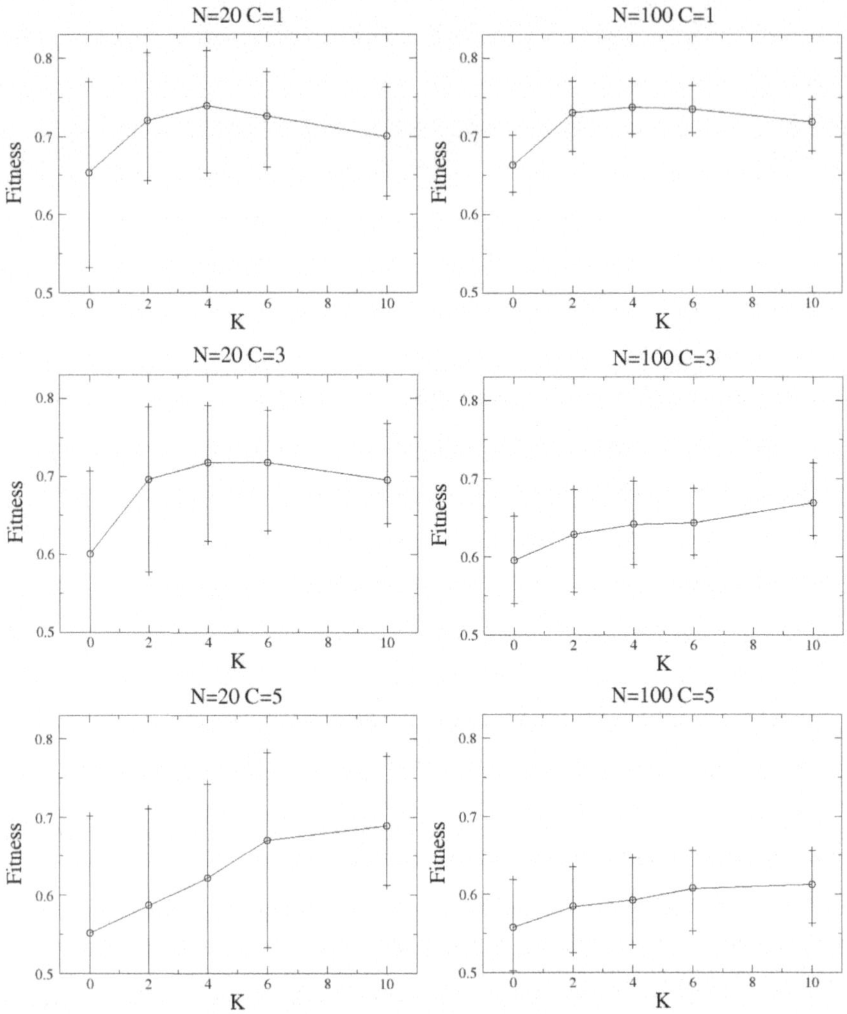

Fig. 4.16 Showing the fitness reached after 20,000 generations for a sexual diploid species coevolving with an asexual haploid species on landscapes of varying ruggedness (K), coupling (C), and length (N)

asexual haploid. Comparing to Fig. 3.2, with $C = 1$ and $N = 20$ or $N = 100$, an increase in fitness compared to the equivalent asexual haploid can be seen for $K > 0$ (T-test, $p < 0.05$), presumably due to the Baldwin effect again being exploited as in the NK model (where $C = 0$). When $C = 3$ and $N = 20$, sexual diploidy results in an increase in fitness over the haploid when $K > 4$ (T-test, $p < 0.05$) but it is always worse with $N = 100$ (T-test, $p < 0.05$). Conversely, with $C = 5$ and $N = 20$, sexual diploidy results in a lower fitness compared to the haploid when $K < 6$ (T-test, $p < 0.05$), with no significant difference for all K when $N = 100$ (T-test, $p \geq 0.05$).

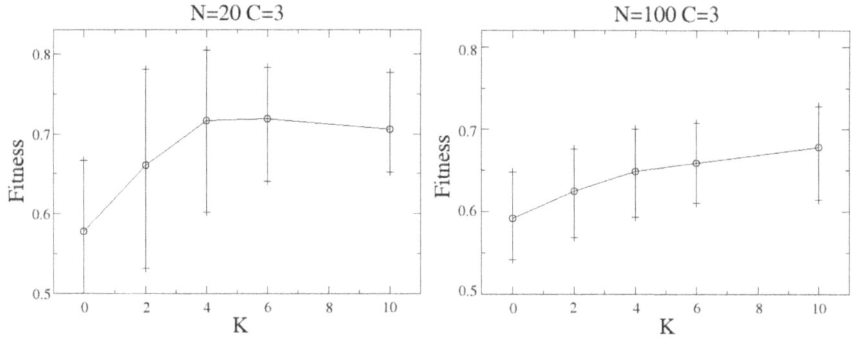

Fig. 4.17 Showing examples of the fitness reached after 20,000 generations for a sexual diploid species coevolving with another sexual diploid species on landscapes of varying ruggedness (K) and length (N) with $C = 3$

Figure 4.17 shows examples when both species are sexual diploids. Comparing to Fig. 4.16, with $C = 1$ or $C = 5$ and $N = 20$ or $N = 100$, there is no difference in fitness compared to the equivalent case coevolving with an asexual haploid for all K (T-test, $p \geq 0.05$). When $C = 3$ and $N = 20$ there is again no difference for all K (T-test, $p \geq 0.05$) but with $N = 100$ an increase in fitness is seen when $K > 4$ (T-test, $p < 0.05$).

As in the traditional NKCS model, the assumption above is that the two species evolve at the same rate. Following results in Chap. 3, the new parameter R can again be added to the model to represent the relative rate at which one of the two species evolves—by undertaking R rounds of mutation and selection to each round by the other. Figure 4.18 shows results for a sexual diploid species coevolving with an asexual haploid species. As can be seen, with $N = 100$, an increase in R again decreases fitness for all K when $C > 1$. Moreover, the fitness reached for $R > 1$ is typically significantly lower (T-test, $p < 0.05$) than in the equivalent case for

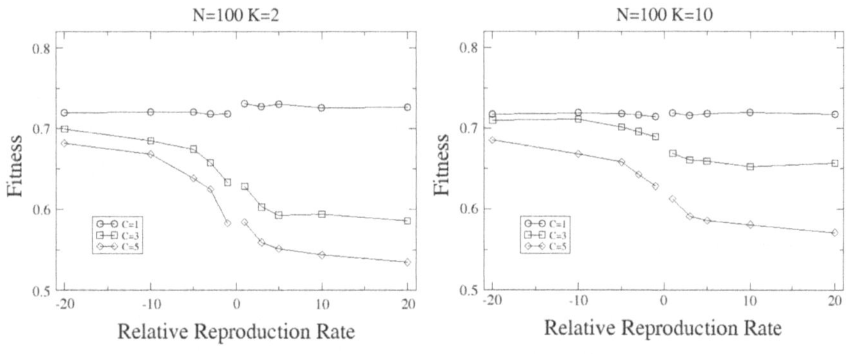

Fig. 4.18 Showing examples of the fitness reached after 20,000 generations for sexual diploid species coevolving with an asexual haploid partner reproducing at a different relative rate R

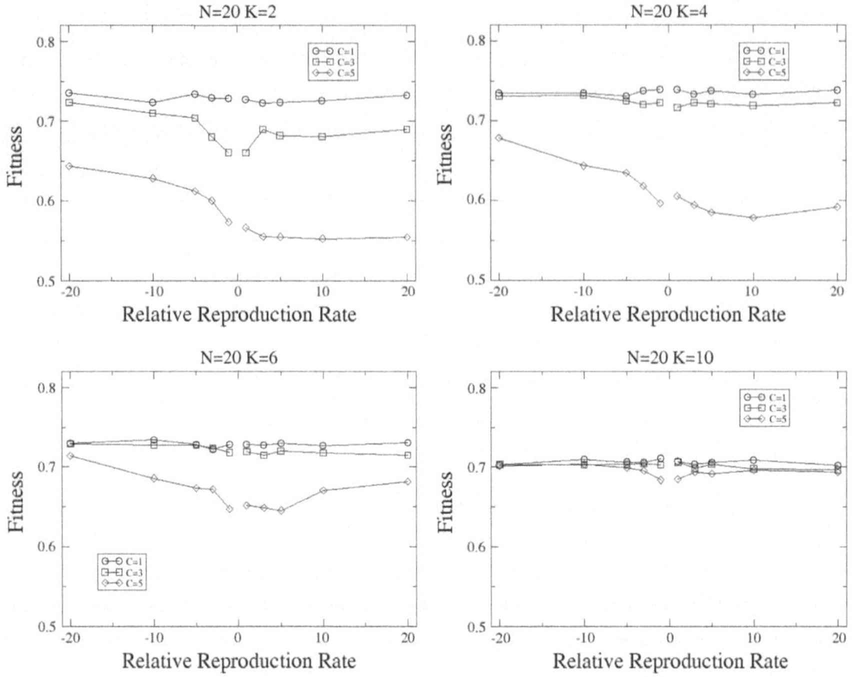

Fig. 4.19 Showing examples of the fitness reached after 20,000 generations for a sexual diploid species coevolving with another sexual diploid species reproducing at a different relative rate R

the asexual haploid species above (compare to Fig. 3.3) when $C > 1$. That is, any advantage (or neutrality) from sexual reproduction seen above with $R = 1$ is lost. Results with $N = 20$ show no significant change in fitness, as was the case with the asexual haploids (not shown).

Figure 4.19 shows the case where both species are sexual diploids. As can be seen, with $N = 20$, there are instances when increasing R can *increase* the fitness of the slower species, when $K = 2$ and $C = 3$, when $K = 6$ and $C = 5$, and when $K = 10$ and $C = 5$ (T-test, $p < 0.05$). This is perhaps somewhat unexpected but again shows how the optimal amount of learning can vary for a given scenario. Results when $N = 100$ are the same as when the partner is asexual above (not shown).

4.6 Discussion

This chapter has suggested that the haploid-diploid cycle seen in all eukaryotic sex exploits a rudimentary form of the Baldwin effect, with the diploid phase seen as the "learning" step [5]. With this explanation for the basic cycle, the other associated phenomena such as recombination, varying the duration of the periods of haploid and diploid state, etc. can be explained as evolution tuning the amount and frequency of learning experienced by an organism. Eukaryotic evolution is seen as refining generalizations over regions of the fitness landscape to identify fit genomes, in contrast to varying the position of single points as in prokaryotic evolution. This explanation does not seemingly contradict any of the mentioned previous explanations for the various stages of eukaryotic sex, rather it presents a unifying process which underpins it and over which many other phenomena may also be occurring.

This hypothesis was based on previous work investigating the Baldwin effect which showed how the optimal amount and/or frequency of learning varied with the ruggedness of the underlying fitness landscape [3]. It is perhaps interesting to note that, in its assuming an animal-like diploid-dominated haploid-diploid lifecycle, conditions exist in the model at $K = 4$ under which both endomitosis and syngamy with a single-step meiosis are equivalent and that syngamy with a two-step meiosis and recombination is most beneficial.

A haploid-diploid cycle has not been shown beneficial in the simplest case of $K = 0$. Some experimental results suggest the average degree of connectivity/epistasis in eukaryotic organisms is typically higher than in prokaryotes (e.g., [14]). This offers one reason why the cycle did not evolve in prokaryotes. Further, it has been suggested in Chap. 3 that the accumulation of mitochondria—and then chloroplasts—through symbiogenesis caused an increase in the ruggedness of the fitness landscape of the resultant early eukaryote as inter-dependence became intra-dependence (after [6]). This can also been seen as creating/aiding the conditions under which a rudimentary Baldwin effect process would prove beneficial. Note that ploidy variation is particularly prevalent in plants (e.g., [19]), where chloroplasts can be seen to further increase K.

Varying ploidy levels in cell types in multicellular organisms can be seen as a further mechanism by which the amount and frequency of learning is fine-tuned. That is, the ruggedness of the fitness landscape contributions for different cell types need not be uniform [9]. The simplicity of the model requires mutational differences between genomes whereas some of the other effects noted above, such as gene product concentrations, could be tuned through varying ploidy levels which may explain why higher ploidy was not beneficial here.

Significantly, it was shown how increasing the size of the genome space ($N = 100$) decreased the benefits of the higher learning rates seen with smaller N ($N = 20$). Thus whilst endomitosis (to diploidy), synagmy, and syngamy with recombination were always found beneficial over the haploid case for $K > 0$, the added benefit of a two-step meiosis with a single recombination point over endomitosis or syngamy was lost with $N = 100$, and a greater number of recombination points were needed to retain the benefit of the latter. However, it is known that "upper and lower tolerance limits for chromosome size seem to exist for some groups of organisms" [18]. The finding here suggests why that is and, moreover, provides a subsequent reason for the maintenance of multiple chromosomes within eukaryotes. Evolution can be seen to tune chromosome length and number to make most effective use of the rudimentary learning process for the overall genome. Since significant increases in a given chromosome's size would disrupt that process, increases in overall genome size are more effectively realized through increasing the number of chromosomes since that can be seen to also increase the number of overall recombination points; dividing genes into chromosomes introduces fixed recombination points in the overall genome, in addition to varying the number of potential recombination events within each of the different chromosomes. This is returned to in the next chapter.

Following results from Chap. 2 where varying the size of genomes during evolution was explored for asexual haploids, results here suggest very similar dynamics occur for sexual diploids. That is, fitness landscape ruggedness does not hinder genome growth and can actually promote it, including enabling the avoidance of the complexity catastrophe for higher levels of gene epistasis for small N. In particular, sexual diploid organisms which experience low levels of sequence deletion compared to that of addition have been found to accept the most novel sequences when they are small. Retrotransposons can perhaps be seen to fulfil this role in eukaryotes.

The coevoloutionary behaviour of sexual diploids has also been explored and they are found to be more sensitive to the effects of fitness landscape coupling than asexual haploids. That is, the Baldwin effect smoothing process is seen to be both beneficial and detrimental on such moving fitness landscapes. These findings extend those above and previously reported in the NK model.

References

1. Bernstein, H., Hopf, F., Michod, R.E.: Is meiotic recombination an adaptation for repairing DNA, producing genetic variation, or both? In: Michod, R.E., Levin, B.R. (eds.) Evolution of Sex: An Examination of Current Ideas, pp. 106–125. Sinauer, Sunderland, MA (1988)
2. Bernstein, H., Bernstein, C.: Evolutionary origin of recombination during meiosis. Bioscience **60**, 498–505 (2010)
3. Bull, L.: On the Baldwin effect. Artif. Life **5**(3), 241–246 (1999)
4. Bull, L.: On coevolutionary genetic algorithms. Soft. Comput. **5**(3), 201–207 (2001)
5. Bull, L.: The evolution of sex through the Baldwin effect. Artif. Life **23**(4), 481–492 (2017)
6. Bull, L., Fogarty, T.C.: Artificial symbiogenesis. Artif. Life **2**(3), 269–292 (1996)
7. Chen, Z.J., Ni, Z.: Mechanisms of genomic rearrangements and gene expression changes in plant polyploids. BioEssays **28**, 240–252 (2006)
8. Cleveland, L.: The origin and evolution of meiosis. Science **105**, 287–289 (1947)
9. Gregory, T.R.: The Evolution of the Genome. Elsevier Academic, Burlington, MA (2005)
10. Haig, D., Grafen, A.: Genetic scrambling as a defence against meiotic drive. J. Theor. Biol. **153**, 531–558 (1991)
11. Hamilton, W.D.: Sex verus non-sex versus parasite. Oikos **35**, 282–290 (1980)
12. Kimura, M.: The Neutral Theory of Molecular Evolution. Cambridge Press, Cambridge (1983)
13. Kondrashov, A.S.: Selection against harmful mutations in large sexual and asexual populations. Genet. Res. **40**, 325–332 (1982)
14. Leclerc, R.: Survival of the sparsest. Mol. Syst. Biol. **4**, 213–216 (2008)
15. Margulis, L., Sagan, D.: Origins of Sex: Three Billion Years Recombination. Yale University Press, New Haven (1986)
16. Maynard Smith, J., Szathmary, E.: The Major Transitions in Evolution. WH Freeman, Oxford (1995)
17. Otto, S.: The evolutionary consequences of polyploidy. Cell **131**, 452–462 (2007)
18. Schlubert, I.: Chromosome evolution. Curr. Opin. Plant Biol. **10**, 109–115 (2007)

19. Soltis, D.E., Visger, C.J., Soltis, P.S.: The polyploidy revolution then and now: Stebbins revisited. Am. J. Bot. **101**, 1057–1078 (2014)
20. Sznajder, B., Sabelis, M.W., Egas, M.: How adaptive learning affects evolution: reviewing theory on the Baldwin effect. Evol. Biol. **39**, 301–310 (2012)

Chapter 5
Chromosomes

Chromosome size, number and types are some of the degrees of freedom exploited by evolution in the variation of eukaryotic organism complexity and this can be seen to have increased over time in some lineages. This chapter explores the effects of fitness landscape ruggedness upon various aspects of chromosomes drawing upon the results in Chap. 4: the effects of varying the number of chromosomes in a sexual diploid are explored using versions of the NK model. Results suggest that landscape ruggedness, chromosome length, and the initial function of the chromosome copy, can all affect evolution. The effects of sex determination chromosomes and hence mating types is also explored within the model, with the XY (ZW) and X0 (Z0) systems shown to be beneficial under certain conditions due to the ruggedness of the fitness landscape.

Dominance has typically been explained either as a consequence of enzymatic pathways with selection playing little or no role (e.g., [5]) or as a consequence of maintained periods of high degrees of allele heterogeneity (e.g., [3]). Under the aforementioned new view of eukaryotic evolution, a new explanation for the emergence of dominance is also presented.

5.1 Chromosome Number

As noted in Chap. 4, since recombination can be viewed as both a mechanism through which the size of the generalizations over the fitness landscape are altered and through which genetic assimilation occurs, the most beneficial rate will probably be influenced by landscape ruggedness (K) and genome size (N). Indeed, it was demonstrated in Fig. 4.8 that fitness could be increased for some values of K ($6 \leq K \leq 15$) with $N =$ 100 using a recombination process where each gene in the non-sister genome copies was swapped with equal probability rather than at a chosen single point. This was suggested as a possible explanation for the variation in the number of types of chromosome seen in eukaryotes since each chromosome may be seen to represent fixed

L. Bull, *The Evolution of Complexity*, Emergence, Complexity and Computation 37,
https://doi.org/10.1007/978-3-030-40730-8_5

recombination points—or fixed points of possible gene mixing—within the overall genome during meiosis. That it may be beneficial for some N and K combinations to have the overall genome subdivided into a number (n) of types of chromosome has been further explored here.

Figure 5.1 shows examples of how fitness can vary with the original pair of haploid genomes of length N subdivided into n equal sections, where $1 \leq n \leq 10$, i.e., there are $2n$ sections in total per diploid. Here each subdivision of N/n genes of the overall N genes undergoes the same process of two-step meiosis with recombination as described above for the original genomes. That is, the first pair of N/n genes is copied and non-sisters recombined, with one of the resulting four sets of genes then chosen at random. This process is repeated n times for each successive section of N/n genes. Once the resulting overall genome is created, mutation is applied as before. With $N = 20$, a significant increase in fitness is found over $n = 1$ when $K = 15$ and $n = 5$ (T-test, $p < 0.10$). With $N = 100$, a significant increase in fitness is found over $n = 1$ when $K = 2$ and $n = 5$ (T-test, $p < 0.05$). Thus it appears dividing a genome into chromosomes is more disruptive than simply increasing the number of recombination points in a single genome.

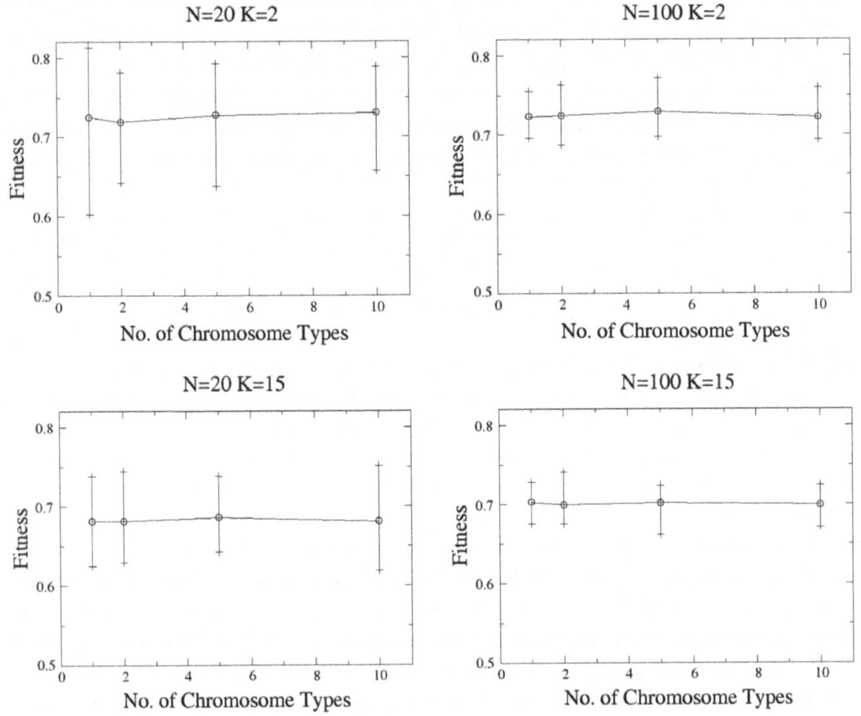

Fig. 5.1 Showing the typical fitness reached after 20,000 generations on landscapes of varying ruggedness where the overall genome (N) has been subdivided into a number of pairs of chromosome types (n)

Whilst it is possible for chromosomes to break or fuse during reproduction, one of the primary mechanisms for the evolution of increased complexity in eukaryotes has been the production and subsequent maintenance of extra copies of one or more chromosomes. Allowing evolution to vary the number of chromosomes of a diploid in this way can be explored within the model.

To enable the number of chromosomes in the model to vary under evolution, i.e., to vary n from an initial $n = 1$ thereby altering the overall genome length, reproduction is here expanded such that either two-step meiosis with recombination occurs for each chromosome pair as described above or a change in the number of chromosomes in one of the two constituent haploid genomes is made. If a change in n is made, a randomly chosen chromosome in a randomly chosen haploid genome is either copied and added to the genome or the last added is removed (equal probability). Deletions and additions are bounded $0 < n \leq 20$, two-step meiosis with recombination only occurs when a corresponding pair of chromosomes exist, with the extra chromosomes of a longer haploid being copied probabilistically half of the time. Finally, mutation is applied to one randomly chosen gene, in one randomly chosen chromosome, in each haploid to create an offspring. The NK model now contains an array of the (maximum) n fitness functions each for N genes for evaluation and overall fitness values are the normalised sum of the given nN gene contributions per haploid. Selection is biased against an increase in n since ties in fitness are decided in favour of the smallest average n, with further ties broken at random as before.

When the n fitness functions are created at random, i.e., as in the standard NK model, no significant increase in chromosome number from $n \approx 1$ is seen and fitness levels remain unaffected for all K, regardless of N (not shown). To some extent this finding is predicted by the results shown in Fig. 5.1 where $n = 1$ is typically as good as the other (pre-determined) values explored. Note the potential for explicit cross-chromosome epistasis in Fig. 5.1 does not exist here either.

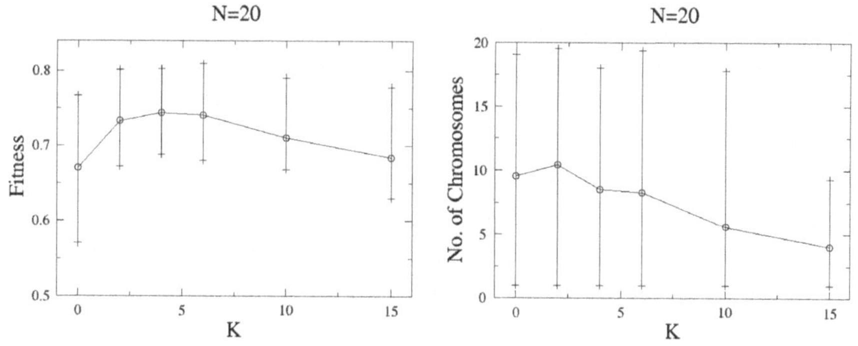

Fig. 5.2 Showing the typical fitness reached after 20,000 generations on landscapes of varying ruggedness and the corresponding average number of chromosomes to emerge from $n = 1$ where all chromosomes are evaluated on the same function

Figure 5.2 shows example results where all n fitness functions are the same. That is, the process of chromosome duplication is said to result in the new copy maintaining the original fitness contribution without disruption, e.g., from dosage effects. Although not explored here, such extra copies are then able to diverge in function over time (e.g., [4]). As can be seen, with $N = 20$, the number of chromosomes which emerge decreases with K, over the range $4 < n < 10$. Moreover, there is no significant change in the fitness reached in comparison to the equivalent fixed $n = 1$ case above, for the majority of K (T-test, $p \geq 0.05$). That is, the duplications are effectively selectively neutral which creates the potential for subsequent divergence in function(s). There is one exception to this finding with $K = 6$ where an increase in fitness is seen (T-test, $p < 0.05$) and $n \approx 8$. In contrast, with $N = 100$, fitness tends to be significantly lower for all K and high average values of n ($n \approx 20$) emerge (not shown).

Although functional divergence is not possible here, it is possible for the copies of the chromosome in an individual to diverge genetically through the accumulation of mutations; the n copies need not be homozygotes over time. Under such circumstances, an individual haploid genome would itself represent two or more points in the underlying haploid fitness landscape, with an average fitness applied to all points here. There is therefore the potential for another fitness landscape smoothing effect from such extra information which appears beneficial around $K = 6$ with $N = 20$ here, but either neutral or detrimental generally.

5.2 Sex Chromosomes

The emergence of isogamy, i.e., mating types, was not considered in the explanation for the evolution of two-step meiosis with recombination in Chap. 4. However, the presence of allosomes—XY in animals and ZW in birds, some fish, reptiles, insects, etc.—can also be explained as a mechanism by which a haploid genome may vary the amount of learning it experiences when paired with another to form a diploid organism. Importantly, taking the view of the constituent haploid genomes, the presence of an heterogametic sex creates the situation where, as evolution converges upon optima, a given haploid containing the common (X or Z) allosome will typically experience two different fitness values simultaneously within a population due to genetic differences between the two sexes; two fitness contributions from the common allosome will almost always exist with two mating types. It is here proposed that the extra (approximate) fitness value information can prove beneficial to the learning/generalisation process described above by adding further landscape smoothing.

To introduce autosomes and allosomes the original pair of haploid genomes of length N are each subdivided into $n = 2$ equally sized chromosomes, i.e., there are $2n$ chromosomes per diploid. A (converged) sub-population of a homogametic sex is said to exist along with a (converged) sub-population of a heterogametic sex. No functional differentiation is imposed upon the heterogametic sex fitness function here; the fitness landscapes of both sexes are identical.

Autosomes undergo two-step meiosis with recombination, as above, whereas allosomes do not undergo recombination. The sex of the offspring is determined by which allosome is (randomly) selected from the heterogametic sex. Once the resulting overall diploid genome is created, mutation is applied to each haploid as before. The fitness contribution of the haploid genomes is their average, as above. For example, when X-inactivation occurs in mammals the choice is typically random per cell lineage in the placenta and hence the fitness contribution of the allosomes remains a composite of the two chromosomes.

Figure 5.3 shows examples of how the benefits of sex chromosomes can vary with landscape ruggedness and size. Note the average fitness of the heterogametic and homogametic sexes is shown. An equivalent (converged) population of hermaphrodites is used for comparison here, where recombination is not used for the second chromosome. As can be seen, for $N = 20$, two sexes prove beneficial for $K > 4$ (T-test, $p < 0.05$) over the hermaphrodite. In contrast, with $N = 100$, two sexes prove detrimental under the same conditions (T-test, $p < 0.05$). Therefore the improvement in fitness seen for sex cannot simply be attributed to the presence

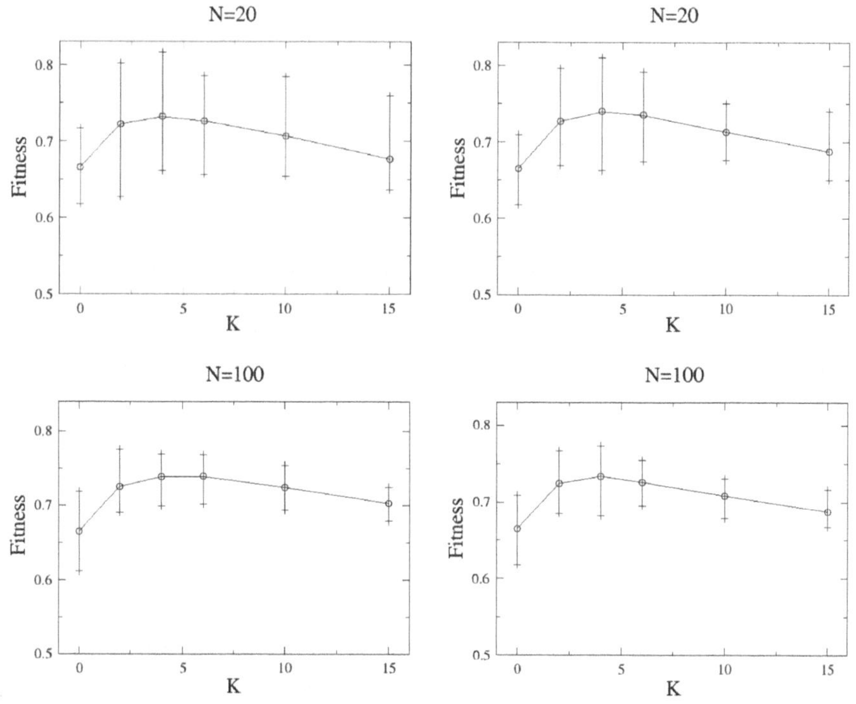

Fig. 5.3 Showing typical behaviour and fitness reached after 20,000 generations on landscapes of varying ruggedness (K) and size (N) for hermaphrodite diploids (left column) or with two sexes (right column)

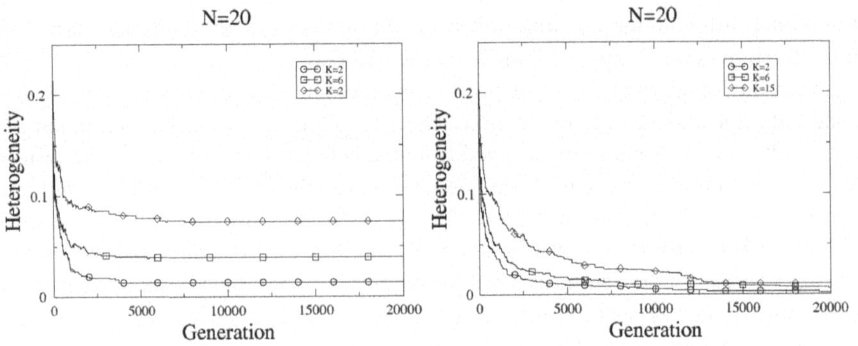

Fig. 5.4 Showing typical convergence behaviour as a fraction of the difference in corresponding gene values in the two haploid genomes ($n = 1$) on landscapes of varying ruggedness (K) for diploids undergoing one-step meiosis (left) and two-step meiosis with recombination (right)

of two explicit sub-populations. There is no significant difference in fitness seen for $K < 6$ (T-test, $p \geq 0.05$) for either N. It can also be noted that including recombination in the second chromosomes (allosomes) of the hermaphrodites does not significantly affect their fitness under any conditions explored (not shown, T-test, $p \geq 0.05$).

As discussed in Chap. 4, increasing the amount of learning under a standard Baldwin effect scheme gives increasing benefit for $K > 6$ with $N = 20$, whereas increasing the amount of learning decreases fitness for $K > 4$ with $N = 100$. As noted above, it is here suggested that the presence of sex chromosomes creates another mechanism through which the fitness landscape of the constituent haploids with the common allosome is potentially smoothed. That is, as the degree of heterogeneity between the two haploids in the homogenetic sex converges (Fig. 5.4, right), two different fitness values/contributions of the chromosome are maintained due to the fitness obtained within the heterogametic sex. This is in contrast to hermaphrodites where eventual near convergence of the two haploids reduces the amount of learning/smoothing as an optimum is found. Therefore the results in Chap. 4 predict those seen here with sex chromosomes added almost exactly: the extra learning mechanism is useful for higher K when $N = 20$ and detrimental for higher K when $N = 100$. Although the extra smoothing proves beneficial on less rugged landscapes ($K > 4$) than in the standard case with $N = 20$, perhaps due to the convergence of the two haploids over time reducing the amount of learning experienced from their partnering.

It can be noted that, whilst varying between reproducing with a member of the opposite sex and as a hermaphrodite is seemingly relatively common in nature (e.g., see [1]), no significant change in behaviour was seen here for a variety of ratios for the parameters explored (not shown). However, the explanation presented that either form of reproduction provides a difference in the amount of learning exhibited suggests a selective advantage for species able to suitably tune the ratio between the two based upon current conditions.

The typical near complete suppression of one X chromosome in females can be seen as a case of extreme dominance where dominance is a common phenomenon in eukaryotes. However, as noted above, explanations for its existence vary. This is explored below in the model.

5.3 Dominance

It was suggested in [2] that dominance can be explained as part of the Baldwin effect view of the evolution of eukaryotes since it can tune the amount of learning experienced on a per-gene basis. That is, dominance can be seen as a mechanism through which evolution is able to bias the composite fitness value assigned to the generalization over the region of the fitness landscape defined by the two constituent haploid genome end points. Hence it can be expected that the fraction of genes with recessive alleles to emerge will vary with the ruggedness of the underlying fitness landscape—the more rugged the landscape, the less dominance (and more learning) expected.

The basic NK model of two-step meiosis with recombination in a diploid (with one type of chromosome, $n = 1$) can be extended to consider a simple dominance mechanism. Here an extra template of length N is added to an individual where each locus can take one of three values—0, 1, #. The first two equate to the dominant allele value for that locus in the case of the two genomes being heterozygote and the last is said to indicate a lack of dominance and hence the fitness contribution is the average of the two genes, as used above. After meiosis, the offspring also has the dominance value of one randomly chosen locus in the template altered to another value, much like the standard gene mutation process. In the case of fitness ties between the offspring and parent, the one with fewer dominated alleles (most #s) is chosen, with subsequent ties broken at random as before. In this way, there is a slight selective bias against dominance. All other details remain as before and individuals are initialised without dominance, i.e., N #s in the template.

Figure 5.5 shows how with $N = 20$ fitness is significantly increased for $K > 0$ and with $N = 100$ fitness is increased for $0 < K < 15$ (T-test, $p < 0.05$, compare to Fig. 4.7). The biggest benefit is seen at $K = 2$ in both cases where the most use of dominated genes emerges, dominance reducing with increasing K thereafter. Hence the fitness bias mechanism for generalizations appears more useful on more correlated fitness landscapes, apart from in the unimodal case of $K = 0$. It can be noted that this general result is also seen when the genomes are initialised as homozygotes (not shown).

5.4 Discussion

Probably through their loss of a rigid cell wall, eukaryotes evolved to exploit DNA arranged in chromosomes. Thereafter chromosome size and number has varied through the processes of duplication and divergence (e.g., [4]). This chapter has used the NK model to explore how fitness landscape ruggedness can affect the evolution

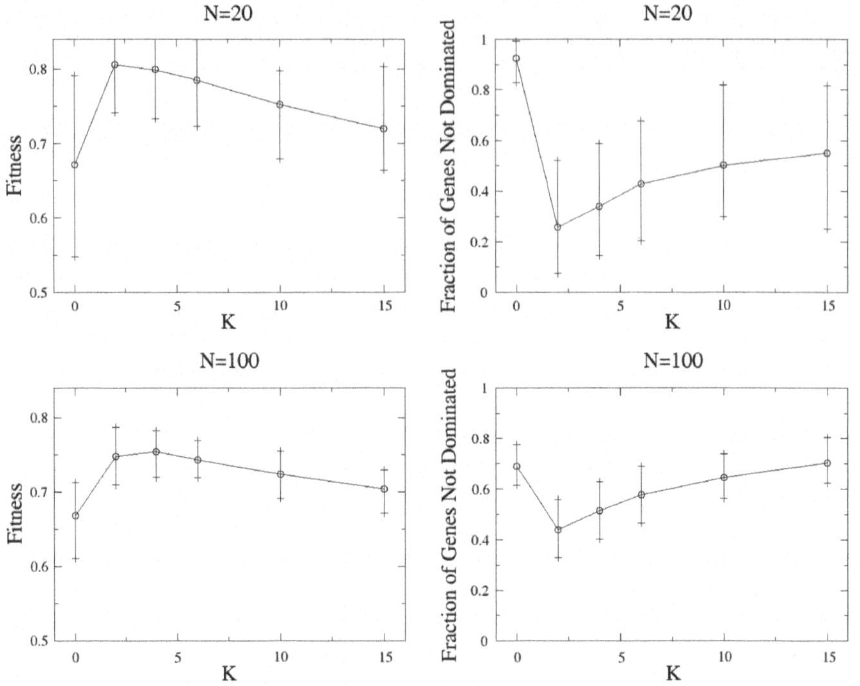

Fig. 5.5 Showing typical fitness (left) reached after 20,000 generations on landscapes of varying ruggedness and size for two-step meiosis with recombination including a dominance mechanism (right)

of such complexity. It has been shown that chromosome number is more likely to increase if the duplicated DNA's initial function remains unaltered, i.e., is selectively neutral, and the landscape is of low ruggedness, and is less likely to prove beneficial as chromosome length increases. This is in contrast to Chap. 4 (and Chap. 2 for asexual haploids) which explored the effects of landscape ruggedness on increases in chromosome length via the addition of novel sequences of DNA where relatively large increases were seen to be beneficial for both small and large chromosomes. Hence the two general mechanisms can be seen to play different roles within the evolution of complexity of eukaryotes.

A number of explanations for the evolution of mating types have been presented (see [1] for an overview). The evolution of allosomes under isogamy was explored here and a beneficial fitness landscape smoothing effect from the extra fitness value information was found for smaller, more rugged landscapes. Since no functional differentiation or recombination was used for the allosomes, the results here apply equally well to the emergence of XY and ZW systems due to the symmetry in the model. Moreover, the same reasoning above applies equally well to the emergence of X0 and Z0 systems, although the effect from the extra fitness value would potentially reduce over time as the homogametic sex converges, depending upon any dosage effects, etc.

Although not explored here, a potential benefit of the XY (ZW) system over X0 (Z0) exists when the diploid fitness landscape is considered. As shown in Fig. 5.4 (right), hermaphrodites exploiting two-step meiosis with recombination can be expected to eventually converge upon organisms carrying two copies of the same (or very nearly) haploid genome. This means that over evolutionary time they become increasingly restricted to the region of symmetry within the diploid fitness landscape, i.e., the region where constituent haploid genome A is genetically similar to haploid genome B. However, the highest levels of fitness may exist in other areas of the diploid landscape, as shown in Fig. 5.6. One way to avoid a near complete set of homozygotes is to maintain a region(s) in the haploid genome where recombina-

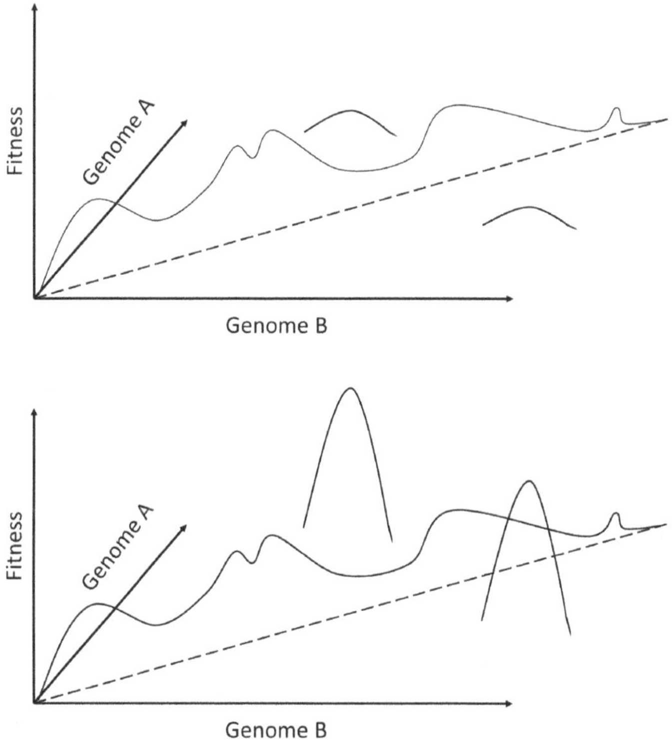

Fig. 5.6 Showing simple example fitness landscapes where sex chromosomes are not (top) and are expected (bottom) to prove beneficial. In the second case, high optima exist away from the area of genome symmetry within the overall diploid fitness landscape created by the two haploids. In the XY (ZW) case, the highest optimum is where both females (males) have genomes corresponding to the mid-point on the axis of possible gene combinations. The existence of sex chromosomes enables the males (females) to maintain genomes corresponding to the end-point on the axis of possible gene combinations and therefore occupy the outlier optima

tion does not occur, thereby inhibiting the assimilation process. It is here suggested this is what the XY (ZW) system enables over the X0 (Z0) system. Asexual reproduction and one-step meiosis, i.e., a haploid-diploid cycle without recombination, can also hinder convergence (Fig. 5.4). Whilst the results in Chap. 4 suggest two-step meiosis with recombination is generally beneficial over asexuality and one-step meiosis due to the increased amount of learning, there is some overlap at low levels of landscape ruggedness (e.g., $K = 2$, in Figs. 4.6 and 4.7). Parabasalid are known to exploit simple syngamy and since they have lost their mitochondria may be seen to exist on less rugged landscapes (after Chap. 3). It can be speculated that they may therefore be exploiting the potential to move away from the region of symmetry within their diploid fitness landscapes. Having adopted the generally more beneficial two-step meiosis with recombination, the XY (ZW) system enables eukaryotes to maintain heterozygotes for some regions of their overall genome space in a relatively controlled manner.

This explains such things as why recombination does not typically occur between allosomes but why recombination over some regions is sometimes seen; the degree of difference is controllable. Relatedly, the two sex chromosomes can be of different sizes which is also potentially correlated to the degree of heterogeneity required between the two haploids to reach the higher fitness areas. Note the situation can be reversed from the XY system in the ZW system since which mating type maintains the single copy of the common allosome is not important in exploiting the benefits described. It also helps to explain why more than two mating types exist in some species—the presence of more than one high fitness region away from the area of symmetry in the fitness landscape is exploitable by the maintenance of a corresponding mating type per region. As well as serving as a mechanism through which the underlying haploid landscape is smoothed through "appropriate" genome pairings, as noted in Chap. 4, sexual selection can also be seen to aid the identification of the higher optima away from the region of symmetry when many exist. Such high optima existing only through the existence of sexual dimorphism would explain the maintenance of sex in even the most unchanging environments. Conversely, that the existence of such regions may vary temporally could help to explain environmental sex determination mechanisms. Moreover, similar reasoning would seem to apply regardless of the details of the mechanism, e.g., for polygenic sex, cytoplasmic control, etc.

Finally, a new explanation for the emergence of dominance has been presented based on the explanation for the evolution of sex in eukaryotes which draws upon the Baldwin effect. That is, dominance is a further mechanism by which the amount of learning experienced may be varied.

References

1. Bachtrog, D., Mank, J.E., Peichel, C.L., Kirkpatrick, M., Otto, S.P., Ashman, T., Hahn, M., Kitano, J., Mayrose, I., Ming, R., Perrin, N., Ross, L., Valenzuela, N., Vamosi, J.C.: Sex determination: why so many ways of doing it? PLoS Biol. **12**(7), e1001899 (2014)
2. Bull, L.: The evolution of sex through the Baldwin effect. Artif. Life **23**(4), 481–492 (2017)
3. Clarke, B.: Frequency-dependent selection for the dominance of rare polymorphic genes. Evolution **8**, 364–369 (1964)
4. Ohno, S.: Evolution by Gene Duplication. Springer, New York (1970)
5. Wright, S.: Fisher's theory of dominance. Am. Nat. **63**, 274–279 (1929)

Chapter 6
Multicellularity

Approximately 550 million years ago the Cambrian explosion brought forth all the major phyla of multicellular animals. Multicellularity is thought to have evolved up to 200 million years before that and has occurred at least three times—in fungi, plants and animals (see [9] for an overview). This chapter explores the effects of fitness landscape ruggedness upon the evolution of multicellularity in eukaryotic organisms.

Eukaryotic green algae range from single-celled organisms (eg, Chlamydomonas) to aggregates of a few cells (eg, of the genus Gonium) to fully multicellular organisms with differentiation (eg, Volvox), prompting the suggestion that multicellularity may have evolved from unicellular aggregates (eg, [7]). This is explored within the NK model and results suggest that genome size, simple differentiation to non-reproduction, as well as further functional differentiation via epigenetic control can all affect the evolution of multicellularity. Significantly, multicellularity is seen to emerge across the parameter space of the model, with an undifferentiated scenario, as might occur in small aggregates, proving most robust before functional differentiation emerges. Such phenomena are considered to be varying the amount of learning in the Baldwin effect.

The correspondences between multicellular organisms and eusocial colonies have long been noted [11]: somatic cells are analogous to the non-reproducing individuals. Indeed, kin-selection [5] has been extrapolated to multicellularity [7]. The findings from exploring multicellularity are consider in the context of eusociality with a number of further similarities identified based on the Baldwin effect.

6.1 Multicellularity in the NK Model

Multicellular organisms are formed by a number of binding mechanisms: in higher plants, the cells are connected via cytoplasmic bridges and exist within a rigid honeycomb of cellulose chambers; and the cells of most animals are bound together by a relatively loose meshwork of large extracellular organic molecules (the extracellular

L. Bull, *The Evolution of Complexity*, Emergence, Complexity and Computation 37,
https://doi.org/10.1007/978-3-030-40730-8_6

matrix) and with adhesion between their plasma membranes. In all cases the cells exist as a larger whole during their lifetime. To take this into account within the sexual diploid NK model described in Chap. 4, the average of the cells' fitnesses is assigned to the organism. That is, in the simplest case, a eukaryotic cell is assumed to divide D times to create D daughter cells. A deterministic mutation rate of $1/N$ is applied to each haploid genome of each daughter cell, as under reproduction. The fitness of the original cell is calculated as before, as is that of each daughter cell, with the resulting fitness summed and divided by $D + 1$.

Figure 6.1 shows example behaviour when $D = 1$ for such simple multicellularity with and without differentiation to a propagule. That is, in the former case, if the offspring multicellular organism replaces the parent, the original cell from which the D daughters were created is subject to the reproduction process of the sexual diploid model described above. In the latter case, one of the $D + 1$ cells in the organism is chosen at random to undergo reproduction.

As can be seen, with $N = 20$ the undifferentiated multicellular scheme is better than simple differentiation to a propagule for $0 < K < 10$ (T-test, $p < 0.05$). Indeed there is no benefit from the simple differentiation scheme over remaining a single cell in such cases (Fig. 4.7, right). No benefit from any form of multicellularity is

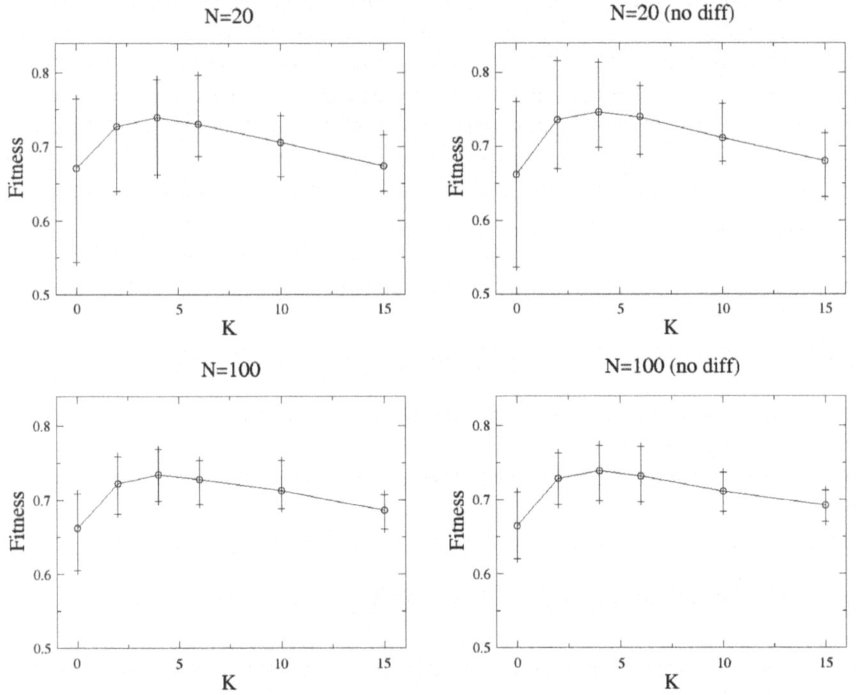

Fig. 6.1 Showing the fitness reached after 20,000 generations on landscapes of varying ruggedness (K) and length (N) with (left) and without (right) simple differentiation to a propagule, $D = 1$

found for all $K > 6$ in comparison to a single cell (T-test, $p \geq 0.05$). For $N = 100$ the benefit from undifferentiated multicellularity over the propagule scheme is lost slightly earlier with it proving fitter when $0 < K < 6$ (T-test, $p < 0.05$) and single cells (Fig. 4.8, left) are fittest thereafter for $6 \leq K \leq 15$ (T-test, $p < 0.05$).

The reason for the benefit from simple multicellularity—and hence for the emergence of multicellularity—can be explained by it representing a further source of learning over the underlying haploid-diploid cycle with meiosis discussed above: the daughter cells represent extra pairs of haploid genome fitness information. Similarity with the explorations of endomitosis in Chap. 4 can be noted here. Chapter 4 demonstrated how higher amounts of learning can prove beneficial for $K > 0$, although that benefit can be lost as either the amount or ruggedness is increased. The simple differentiated form of multicellularity where only the original mother/propagule cell reproduces did not prove beneficial: in contrast to the case where either cell may become the mother cell, too much learning occurs in such cases and the reproducing haploid genomes are too far displaced from the extra information provided by the daughter cell for evolution to be successfully guided to the higher fitness regions.

Figure 6.2 shows examples of the effects from varying the number of daughter cells, where $0 < D \leq 10$. As can be seen, and as suggested by the results above

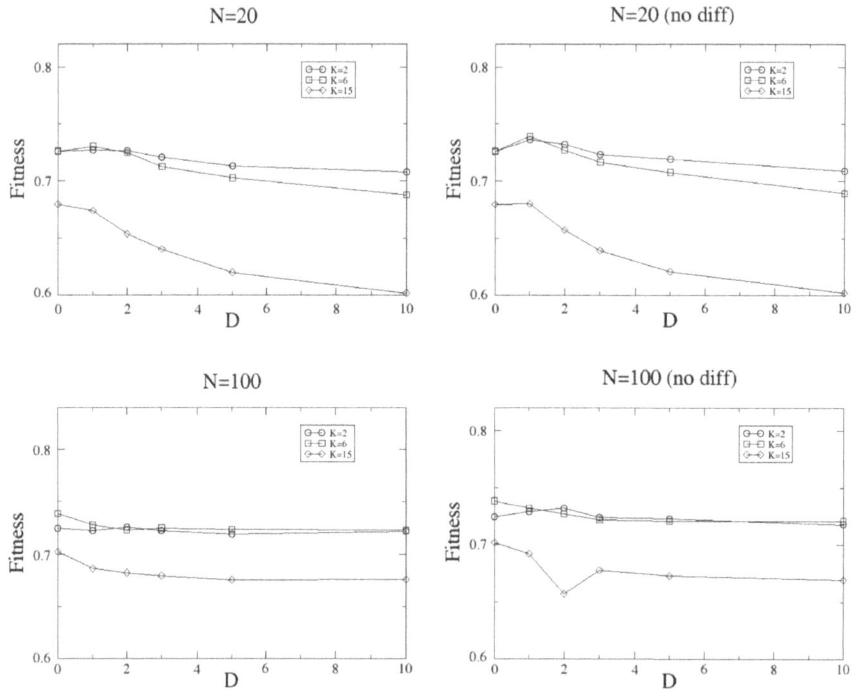

Fig. 6.2 Showing the fitness reached after 0 generations on landscapes of varying ruggedness (K) and size (N) with and without simple differentiation to a propagule for varying numbers of daughter cells (D)

with $D = 1$, the simple differentiation scheme provides no benefit over a single cell ($D = 0$) due to the increased level of learning with increasing D. The undifferentiated scheme does not benefit further from increasing D, although fitness levels remain unaffected when $0 < D < 3$ with $N = 100$ and $2 < K < 10$ (T-test, $p \geq 0.05$). It can be noted that, and as predicted by the above, if the daughter cells are produced in sequence, that is, cell d_i is produced via the mutation of cell d_{i-1} as opposed to the mutation of cell d_0, fitnesses are significantly lower for all $D > 1$ (not shown).

6.2 Functional Differentiation and Simple Epigenetic Control

In the above, the daughter cells were not assumed to be functionally differentiated from the original mother cell—all cells were evaluated on the same fitness landscape. Most models of multicellularity assume functional differentiation in the daughter(s) such as improved feeding ability, eg, see [12], and development in general has been directly linked to the evolution of complexity [1]. Indeed, in contrast to the results above, it is typically some form of division of labour between the cells that is used to explain the emergence of multicellularity. The above model has been altered such that the daughter cells are evaluated on a second (randomly created) NK fitness function of the same size and ruggedness as the mother cell.

Figure 6.3 shows examples of how undifferentiated multicellularity is beneficial over simple differentiation to a propagule for $2 < K < 15$ with $N = 20$ (T-test, $p < 0.05$). However no significant difference in fitness is seen between either form of multicellularity when $N = 100$ (not shown).

Epigenetics refers to cellular mechanisms that affect transcription without altering DNA sequences, with the two principal mechanisms being methylation and histone modification. In the former case, a methyl group attaches to the base cytosine, or

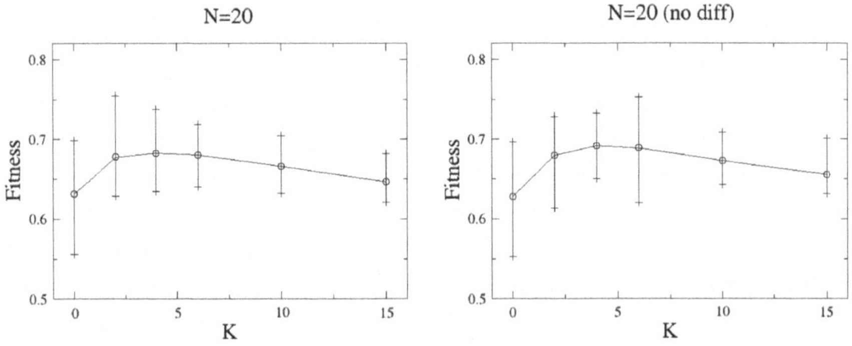

Fig. 6.3 Showing the fitness reached after 20,000 generations on landscapes of varying ruggedness (K) with (left) and without (right) simple differentiation to a propagule, where the daughter cell is evaluated on a different fitness landscape to the mother ($D = 1$)

adenine in bacteria, typically causing a reduction in transcription activity in the area. In the latter case, changes in the shape of the proteins around which DNA wraps itself to form chromatin can alter the level of transcription in the area. The above model can be extended to include simple epigenetic control thereby enabling functional differentiation between the cells.

Similar to the scheme suggested in [10], an extra template of maximum length 2 N is added to an individual where, in the case of $D = 1$, each locus can take one of three values—0, 1, # (Fig. 6.4). The first two equate to which cell the pair of genes at the corresponding locus in the haploid genomes are used and the last is said to indicate a lack of cell specificity and hence the two genes are used in both cells, as above. Each haploid genome is now of variable length N' in the range $[N, 2N]$, with $D = 1$. After meiosis, the offspring also has the epigenetic control value of one randomly chosen locus in the template altered to another value or the template length is randomly increased/decreased by one. In the case of a change of size in the template, the genome lengths are also adjusted and a new (random) gene added onto the end or the last gene added is removed from each haploid (i.e., as in Chap. 2). In the case of fitness ties between the offspring and parent, the shorter genome/template individual is chosen, with subsequent ties broken at random as before. In this way, there is a selective bias for the most efficient way to define $D + 1$ functionally differentiated cells. That is, individuals of length $N' = 2N$, with a template of N 0's and N 1's, i.e., a complete genome specification for each cell type, are discouraged. All other details remain as before and individuals are initialised without differentiation, i.e., $N' = N$ and N#s in the template. Note too few genes for a cell is penalised since fitness contributions are still divided by N. Conversely, only the first N genes are considered when more are specified for a given cell.

Figure 6.5 shows how for both $N = 20$ and $N = 100$, the epigenetic control mechanism enables significantly improved fitness for $K > 0$ (T-test, $p < 0.05$) in comparison to both forms of multicellularity explored above on the heterogeneous NK functions (Fig. 6.3). It can also be seen that cell-specific genes emerge with

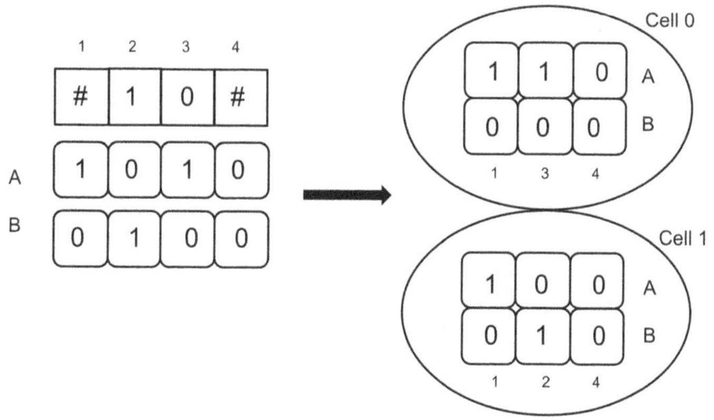

Fig. 6.4 Showing an example of the epigenetic template scheme with $D = 1$

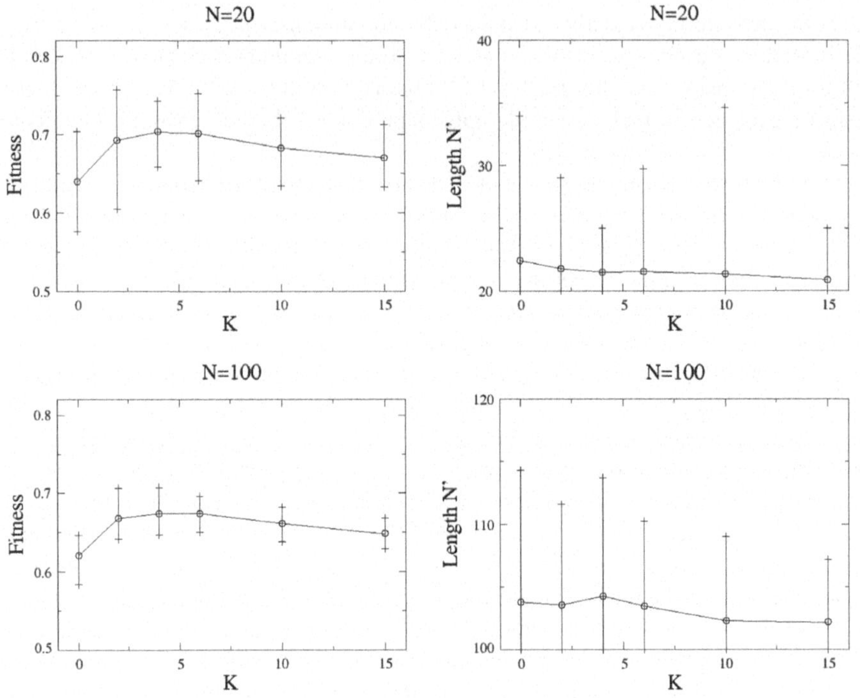

Fig. 6.5 Showing the fitness reached after 20,000 generations (left) on landscapes of varying ruggedness (K) and size (N) with epigenetic control, where the daughter cell is evaluated on a different fitness landscape to the mother ($D = 1$). The average length of the genomes is also shown (right)

$N' > N$ in all cases, with a slight trend for lower N' with increasing K. Fitness levels are not as good as those for the standard single cell case (Fig. 4.7) but that may be due in part to the simple epigenetic control template mechanism and its variation used.

6.3 Eusociality: Haplodiploid Multicellularity

Eusocial species are found amongst termites, ants, wasps, bees, aphids, mole-rats and spiders. Wheeler [11] was first to highlight the close analogy between a eusocial colony and a multicellular organism in that both have differentiation into reproductives and non-reproductive specialists. Hamilton's [5] kin-selection explains why (subsocial) eusocial offspring give up their right to reproduce: a daughter has as many genes in common with its own offspring as its mother's and hence there is no selective difference between it raising its children or siblings. As noted above, Maynard-Smith and Szathmary [7, p8] have suggested Hamilton's theory can be applied to multicellularity, following Wheeler's insight. That is, kin-selection can also be attributed to somatic cells. This contrasts with Buss [3] who postulates propagule control of soma and Michod (eg, [8]) who has highlighted other factors, specifically cell-cell policing and germ-line segregation (reducing the potential for soma conflict), as being equally as important. Policing is also thought to have been important in the emergence of eusocial colonies [4].

The above model can be extended to explore the emergence of a non-reproductive haploid individual from a diploid. Figure 6.6 shows the case where a daughter is created as a mutated copy of the first (arbitrary) haploid genome from the diploid produced under meiosis. The mother is assigned the average fitness of itself and that of its daughter. When the mother and daughter are evaluated on the same fitness landscape (left column) with $N = 20$, eusociality is fitter than the simple multicellular differentiation case of a propagule (Fig. 6.1) when $K > 4$ (T-test, $p < 0.05$) and no different to the undifferentiated scheme apart from being worse when $K = 2$. When $N = 100$, the same is generally true, although eusociality performs the same as the multicellular undifferentiated scheme, including for $K = 2$ but is fitter for $K = 6$

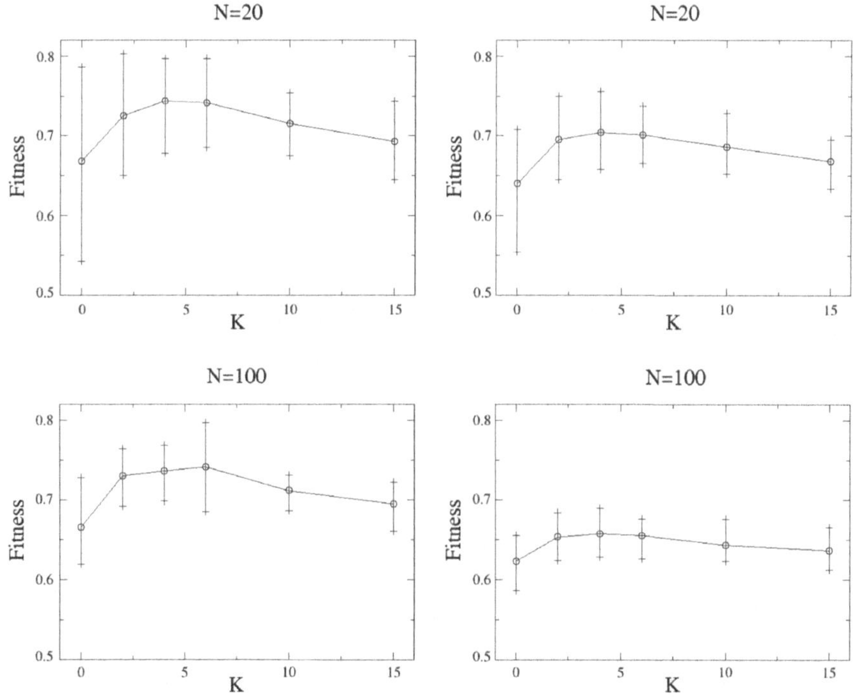

Fig. 6.6 Showing the fitness reached after 20,000 generations on landscapes of varying ruggedness (K) and size (N), where the daughter haploid is evaluated on the same (left) and a different (right) landscape

(T-test, $p < 0.05$). When the mother and daughter are evaluated on two different functions (right column), i.e., as in Fig. 6.3, with $N = 20$, eusocialty is always fitter than either diploid-diploid multicellular scheme when $K > 0$ (T-test, $p < 0.05$) and the same when $N = 100$ for all K (T-test, $p \geq 0.05$). Hence the loss of the effects of the learning within a second diploid, as in multicellularity, reduces the amount of learning overall in eusociality which appears to result in a more appropriate amount for some N and K combinations.

Figure 6.7 shows the effects of including the same epigenetic control template as above into the diploid-haploid eusocial scheme as a form of morphological differentiation in the colony members. As with multicellularity, fitnesses are always improved over the non-epigenetic case (Fig. 6.6, right column) when $K > 0$ (T-test, $p < 0.05$). Such epigenetic/differentiated eusociality is fitter than epigenetic multicellularity when $K > 0$ with $N = 20$ and when $0 < K < 6$ with $N = 100$.

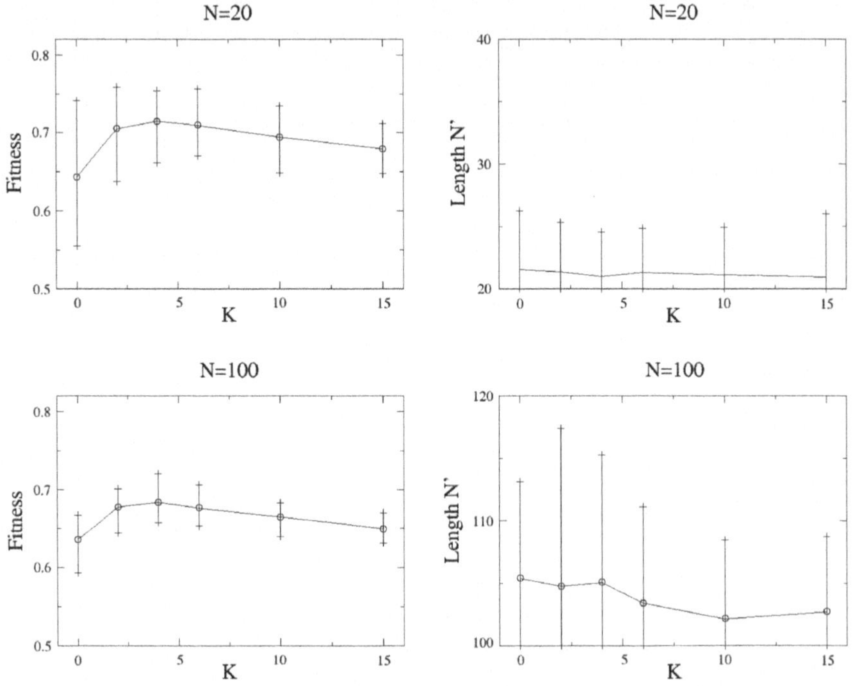

Fig. 6.7 Showing the fitness reached after 20,000 generations (left) on landscapes of varying ruggedness (K) and size (N) with epigenetic control, where the daughter haploid is evaluated on a different fitness landscape to the mother. The average length of the genomes is also shown (right)

6.4 Discussion

This chapter has considered the effects of fitness landscape ruggedness on the evolution of multicellularity in sexual diploids. Using versions of the NK model it has been shown that simple forms of multicellularity can emerge over a significant range of amounts of landscape ruggedness due their further exploiting the Baldwin effect beyond that inherent to the haploid-diploid cycle with meiosis. In particular, the more basic form under which either/any cell in the organism may become the propagule was found to be more widely beneficial. That multicellularity may have emerged through the exploitation of a form of Baldwin effect has been suggested previously, although asexual haploid cells/organisms were assumed in the models [2]. Whilst some bacteria aggregate together to avoid starvation or predation (eg, see [6]), multicellularity is a common eukaryotic phenomenon. The results here suggest the reason why this is: bacteria do not typically exist on sufficiently rugged fitness landscapes to benefit from the Baldwin effect. In Chap. 3 the presence of mitochondria in eukaryotes was suggested as causing an increase in the ruggedness of their fitness landscapes.

The form of multicellularity and the number of daughter cells produced has been shown to vary the amount of learning occurring during evolution, where it was shown how the amount which is beneficial depends upon the fitness landscape size and ruggedness. The simplicity of the model requires mutational differences between genomes in the mother and daughter cell(s) whereas other cellular processes, such as gene product extracellular concentrations, could be tuned through varying the number of cells to similar effect.

Moreover, as with the other sources of innovation considered in this book, no explicit cost for multicellularity was included—but improved fitness was found. Hence if the cost is less than the benefit, multicellularity can be expected emerge. Thereafter functional differentiation can open new niches where competition does not exist; simple epigenetic control has been explored and shown widely beneficial.

Eusociality has emerged in the Hymenoptera (ants, bees, wasps, etc.) multiple times and it is their haplodiploidy that has been the major factor in this (after [5]). The results in this chapter suggest that the Baldwin effect was also significant to the emergence of eusociality—for the same reasons as for multicellularity (after [2]). In particular, this is supported by the fact that the oldest eusocial organisms are the termites which are diploid and so are more similar to the multicellular case. It has been shown that haploid daughters are similarly beneficial to diploid-diploid schemes on rugged fitness landscapes, particularly when simple differentiation to a propagule exists. This can again be explained by the amount of learning occurring beyond that inherent to the sexual diploids. Following the approach in Chap. 4 of considering the haploid genome fitness landscape of primary importance, as opposed to that of the diploid, the daughters under haplodiploidy can be viewed as not using the rudimentary learning process for the genes contained in the diploid mother. That is, the mothers contain two haploid genomes and the fitness contribution of the genes therein will be some combination of the two, eg, their average was used above.

In contrast, in the daughter the fitness contribution of the genes is simply as in the haploid case, i.e., without the potential for any learning since they are not paired with another set of corresponding genes. Thus a daughter can be seen as a mechanism through which the learning experienced in the region of the haploid genome fitness landscape covered by the genes contained in her genome is removed; a daughter provides additional "raw" haploid fitness contribution information to the mother. Similar reasoning was used to explain the emergence of X0 (Z0) sex chromosomes in eukaryotes in Chap. 5.

Eusociality was found to be most beneficial in the cases without epigenetic/morphology control when the mother and daughter exist on different fitness landscapes with smaller genomes. When simple differentiation control was added it also became most beneficial for larger fitness landscapes under some conditions. In all cases, the conditions for the emergence of more complex organisms can again be seen to exist across the parameter space of the model via the use of the Baldwin effect to smooth the underlying rugged fitness landscape.

References

1. Bonner, J.T.: The Evolution of Complexity by Means of Natural Selection. Princeton University Press (1988)
2. Bull, L.: On the evolution of multicellularity and eusociality. Artif. Life **5**(1), 1–15 (1999)
3. Buss, L.W.: The Evolution of Individuality. Princeton University Press (1987)
4. Frank, S.A.: Mutual policing and repression of competition in the evolution of cooperative groups. Nature **377**, 520–522 (1995)
5. Hamilton, W.D.: The genetical evolution of social behaviour. J. Theor. Biol. **7**, 1–52 (1964)
6. Herron, M., Borin, J., Boswell, J., Walker, J., Chen, I., Knox, C., Boyd, M., Rosenzweig, F., Ratcliff, W.: De novo origins of multicellularity in response to predation. Nat. Sci. Rep. **9**, 2328 (2019)
7. Maynard Smith, J., Szathmary, E.: The Major Transitions in Evolution. WH Freeman, Oxford (1995)
8. Michod, R.: Cooperation and conflict in the evolution of individuality. Proc. R. Soc. Lond. B Biol. Sci. **263**, 813–822 (1996)
9. Nedelcu, A., Ruiz-Trillo, I. (Eds.): Evolutionary Transitions to Multicellular Life: Principles and Mechanisms. Springer, Dordrecht (2015)
10. Turner, A., Lones, M., Fuente, L., Stepney, S., Caves, L., Tyrell, A.: The incorporation of epigenetics in artificial gene regulatory networks. BioSystems **112**, 56–62 (2013)
11. Wheeler, W.M.: The ant-colony as an organism. J. Morphol. **22**, 307–325 (1911)
12. Wolpert, L.: The evolution of development. Biol. J. Lin. Soc. **39**, 109–124 (1990)

Chapter 7
Conclusion

> [T]here is no reason why evolution by natural selection should lead to an increase in complexity … an increase in immediate 'fitness' – that is, expected number of offspring – may be achieved by losing eyes or legs as well as by gaining them. [2, p. 4]

Whilst there is no clear correlation, complex organisms typically contain more DNA than simple organisms—whether it is in the genome, due to multiple genomes in a cell, due to the presence of symbiotic organelles, or due to the number of cells. Hence it is an indicative measure of organismal complexity, although others have been proposed (eg, see [1] for discussions). The conditions under which such increases in DNA might emerge during evolution have been explored using simple fitness landscape models here. These models can perhaps be seen as a method by which to attempt to explain biology whilst removing as much detail of the biology as possible, and thereby potentially missing key elements. For example, although touched upon for symbiogenesis, that an increase in complexity may subsequently create the opportunity for conflict has not been explored. However, such abstract fitness landscapes enable the consideration of evolution purely as a stochastic process being fed innovations due to (often almost unavoidable) things like copy errors, joining/separation events, etc. as it searches through a multidimensional space containing peaks and troughs. The findings here show how the ruggedness and movement/change within fitness landscapes can be used to explain the *emergence* of key complexity increasing events.

Chapter 2 showed how the adoption of new DNA with random functionality is seen across the parameter space as evolution searches over rugged fitness landscapes of all types, particularly in the early stages of adaptation, eg, after an environmental change has moved the relative position of optima. Thus such simple increases in complexity appear to be an inherent feature of evolution. It can also be noted that reductions in complexity were seen in response to change under some conditions.

Chapter 3 showed how the bringing together of genomes from separate lineages is almost always beneficial if those organisms interact sufficient closely in their environments, particularly as the genome length of those organisms increases. Thus

L. Bull, *The Evolution of Complexity*, Emergence, Complexity and Computation 37,
https://doi.org/10.1007/978-3-030-40730-8_7

increases in ruggedness and complexity through the formation of close symbiotic relationships again appears to be an inherent feature of evolution. Note the correlation with findings in Chap. 2: the close interactions cause movement in the fitness landscapes of the partners which can encourage increases in genome length. But it must be noted that the seemingly long delay between the emergence of the first cell and the first eukaryotic cell suggests that the probability of that particular innovation was low in natural evolution.

Chapter 4 showed how doubling the number of genomes in a cell to form a diploid is increasingly beneficial as the ruggedness of the fitness landscape increases, with tetraploidy similarly typically beneficial under some conditions. Note the correlation with findings in Chap. 3 where ruggedness is increased: sexual diploids have endosymbiotic organelles. A simple form of the Baldwin effect is suggested as causing this benefit. Moreover, a two-step meiosis with recombination in a haploid-diploid lifecycle appears beneficial over both syngamy and endomitosis as fitness landscape ruggedness increases. Hence sex appears to be a consequence of this increase in complexity. And it was suggested that through sex evolution exploits a form of generalization in eukaryotes.

Chapter 5 showed how less complex organisms will increase the number of (functionally homogeneous) chromosomes in a cell. Thus the potential for future functional divergence/novelty increases. It was also shown how the emergence of two mating types through an example of such chromosome duplicate divergence is beneficial for less complex organisms as fitness landscape ruggedness increases. Note the correlation with findings in Chap. 4: the conditions for the emergence of sex are typically those where mating types emerge.

Finally, Chap. 6 showed how multicellular organisms of just a few cells are either beneficial or selectively neutral on all but the most or least rugged fitness landscapes. Again, a simple form of the Baldwin effect is suggested as the underlying cause of the benefit. Since this is the case with or without differentiation to a non-reproducing propagule, simple multicellularity appears very likely—even before the benefits of full functional differentiation are realised. Note the correlation with findings in Chap. 4: sexual diploids exist on fitness landscapes suitable to exploit such simple multicellularity.

With the caveats expressed above, the findings here suggest that from a purely evolution-as-search-process view, that the complexity of life has been increasing over the last 3.8 billion years is far from surprising due to inherent properties of the process.

References

1. Bedau, M.: The evolution of complexity. In: Barberousse, A., Morange, M., Pradeu, T. (eds.) Mapping the Future of Biology: Evolving Concepts and Theories, pp. 106–125. Springer, New York (2008)
2. Maynard Smith, J., Szathmary, E.: The Major Transitions in Evolution. WH Freeman, Oxford (1995)

Appendix
Regulation

The simple NK and NKCS models used throughout this book can be extended such that particular aspects of gene regulation, which might be enabled by an increase in the amount of DNA in an organism, can be explored in more detail. This appendix provides an example.

A.1 The RBN and RBNK Models

Random Boolean networks (RBN) [7] were introduced as an abstract model by which to explore aspects of genetic regulatory networks. RBN consist of R genetic loci/nodes, each connected to B other randomly chosen nodes, with each performing a randomly assigned Boolean update function based upon the current state of those nodes per update cycle (Fig. A.1). Hence those B nodes are seen to have a regulatory effect upon the given node; the details of transcription are abstracted out.

Since they have a finite number of possible states and they are deterministic, such networks eventually fall into an attractor. It is well-established that the value of B affects the emergent behaviour of RBN wherein attractors typically contain an increasing number of states with increasing B. Three phases of behaviour were originally suggested through observation: ordered when $B = 1$, with attractors consisting of one or a few states; chaotic when $B \geq 3$, with a very large number of states per attractor; and, a critical regime around $B = 2$, where similar states lie on trajectories that tend to neither diverge nor converge (see [8] for discussions of this critical regime, e.g., with respect to perturbations). Subsequent formal analysis using an annealed approximation of behaviour identified $B = 2$ as the critical value of connectivity for behaviour change [5]. Example behaviour is shown in Fig. A.2.

L. Bull, *The Evolution of Complexity*, Emergence, Complexity and Computation 37,
https://doi.org/10.1007/978-3-030-40730-8

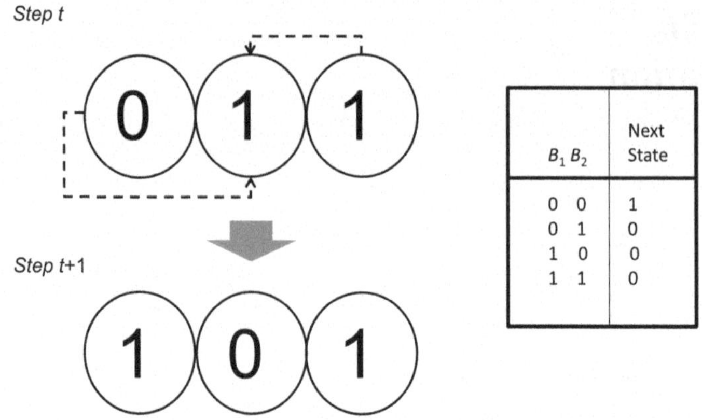

Fig. A.1 An example RBN model (R = 3, B = 2) over one update cycle, where there are 2^B possible regulatory input combinations per node, each of which is assigned a random Boolean function update table (right, centre node shown)

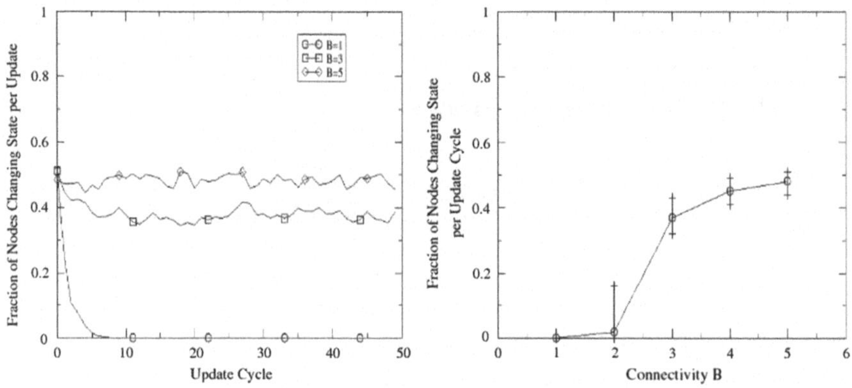

Fig. A.2 Typical behaviour of RBN with R = 100 nodes: on the left, showing example temporal dynamics; and on the right, the average behaviour (100 runs) after 100 update cycles. Nodes were initialized at random

With the aim of enabling systematic exploration of the evolvability of such GRN, RBN have been combined with the NK model of fitness landscapes. In the combined form—termed the RBNK model [1]—a simple relationship between the states of N randomly assigned nodes within an RBN is assumed such that their value is used within a given NK fitness landscape of trait dependencies. Hence the NK element creates a tuneable component to the overall fitness landscape with behaviour (potentially) influenced by the environment though N_i inputs to other nodes (Fig. A.3). Moreover, the conditions under which specific regulatory functions may emerge can be explored.

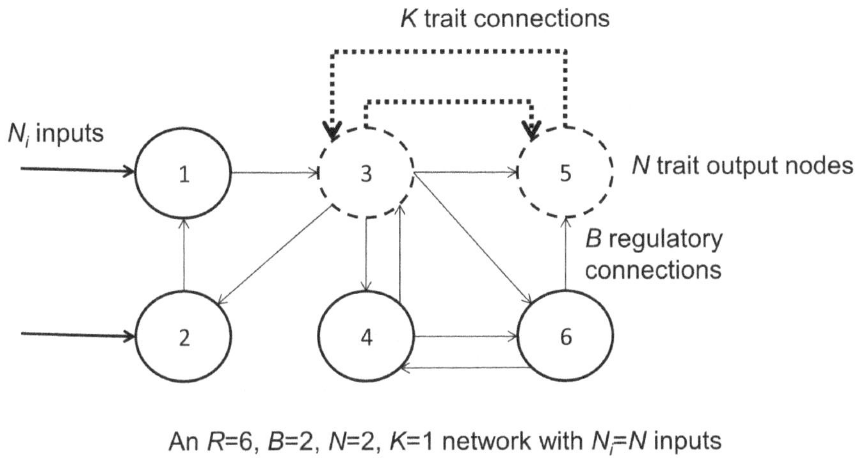

Fig. A.3 The combined RBN and NK modelNK model—the RBNK modelRBNK model

Gene regulatory network complexity is known to have increased over time, e.g., as the driver of metazoan morphological complexity [4]. The RBNK model enables the systematic examination of specific regulatory mechanisms—in a similar way to the NK model is used above—through which such complexity has occurred, such as the role of RNA editing. RNA editing is the alteration of the sequence of RNA molecules through a variety of mechanisms after initial expression (see [9] for an overview). In some cases such editing is triggered by specific conditions, in others it is necessary for the normal function of a cell. RNA editing is a widespread form of regulatory control and appears to have evolved many times (e.g., see [6]). For example, guide RNA (gRNA) are relatively small molecules that align themselves to complementary regions of messenger RNA (mRNA) and either insert or delete a base(s) thereby (typically) altering the structure of the protein specified in the expressed DNA. In this appendix, RBNs are extended to include a simple form of RNA editing [3]. The selection of the extra mechanism is explored under various scenarios, drawing upon the findings of the chapters above regarding the conditions under which the pre-requisite increase in genome length may occur.

A.2 gRNA in the RBNK Model

To include a mechanism which enables the modification of transcription based upon the internal and/or external environment of the cell, nodes in the RBN can (potentially) include a second set of B' connections to defined nodes and another Boolean update function. Each such node also maintains a table containing a list of node id's for each entry in the Boolean table for the B' connections where the node is on/expressed. The list is the same size as the out-degree of the node (range [0, $R*B$]).

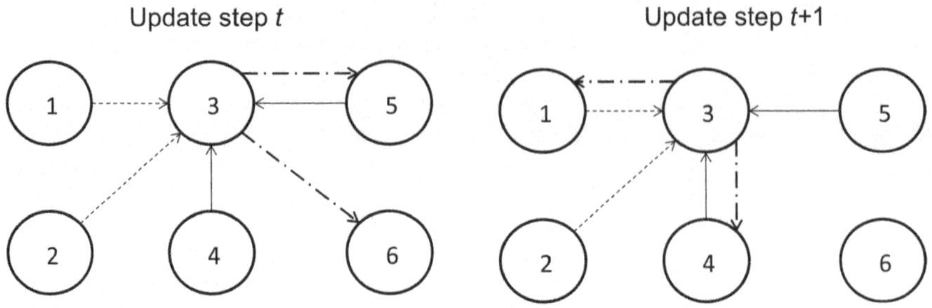

B_1 B_2	Next State	B'_1 B'_2	Next State	Next Output Connections
0 0	1	0 0	1	1 4
0 1	0	0 1	1	5 6
1 0	0	1 0	0	- -
1 1	0	1 1	0	- -

Fig. A.4 Example RBN with RNA editingRNA editing. The look-up table and connections for node 3 only are shown for clarity. If every node in the RBN is in state '0' at time '*t*', node 3 will turn on for the next time step, as will its associated gRNA. As a consequence, nodes 1 and 4 will be connected to node 3 for time step $t + 1$ for their updating

This is seen as introducing a non-coding RNA associated with the protein expressed by the given node. RNA editing causes a change in the connectivity of the RBN which lasts for one update cycle.

On each traditional RBN update cycle, the connectivity of the network is initially assumed (reset) to be that originally determined by evolution. Then, for each node which has an associated guide RNA, a check is made to see if the node's transcription state was set to on ('1') on the last update step and if its associated RNA has been activated ('1') since the last time this occurred. If so, the out-connections for that node are altered to those in the corresponding table entry for the current state of the *B'* connection nodes (Fig. A.4). If a node subsequently has fewer than *B* input connections, the missing gene(s) is simply assumed to not be expressed on that cycle, i.e., the gene on the end of the connection is assumed to be set to 0. If a node subsequently has more than *B* connections, the "extra" input is randomly assigned to one of the existing *B* connections and a logical OR function is used to determine whether that connection is considered to be to an expressed gene. Thereafter each node updates its transcription state based upon the current state of the nodes it is (currently) connected to using the Boolean logic function assigned to it in the standard way, as do any associated non-coding RNA node elements. For simplicity, the number of standard regulatory connections is assumed to be the same as for RNA editing, i.e., $B = B'$.

A converged population of asexual haploids is assumed here, as above Each RBN is represented as a list to define each node's start state, Boolean function for transcription, *B* connection ids, *B*' connection ids, Boolean function for RNA editing, re-connectivity entries under RNA editing, and whether it is an RNA edited node or not. Mutation can therefore either (with equal probability): alter the Boolean transcription function of a randomly chosen node; alter a randomly chosen *B* connection; alter a node start state; turn a node into or out of being RNA editable; alter one of the re-connection entries if it is an editable node; or, alter a randomly chosen *B*' connection, again only if it is an editable node. A single fitness evaluation of a given GRN is ascertained by updating each node for 100 cycles from the genome defined start states. An input string of N_i 0's is applied on every cycle here. At each update cycle, the value of each of the *N* trait nodes in the GRN is used to calculate fitness on the given NK landscape. The final fitness assigned to the GRN is the average over 100 such updates here. A mutated GRN becomes the parent for the next generation if its fitness is higher than that of the original. In the case of fitness ties the number of RNA editable nodes is considered, with the smaller number favoured, the decision being arbitrary upon a further tie. *Hence there is a slight selective pressure againstRNA editing*. Here $R = 100$, $N = 10$ and results are averaged over 100 runs—10 runs on each of 10 landscapes per parameter configuration—for 50,000 generations, $0 < B \leq 5$ and $0 \leq K \leq 5$ are used. As Fig. A.5 (left column) shows, regardless of *K*, RNA editing is selected for in all high connectivity cases on average, i.e., when $B > 3$, when the underlying fitness landscape is unchanging. Analysis of the behaviour of the editing in such cases indicates that it is applied throughout the lifecycle, although a clear a pattern of usage is typically difficult to establish, often with varying numbers of nodes exhibiting editing per cycle.

Since it is known such highly connected networks typically exhibit chaotic dynamics and they are subsequently difficult to evolve [1], it might be surmised that the RNA editing is not performing a functional role, rather it is maintained under drift/neutral processes. As noted above, RNA editing alters the out-connections of a given node and hence a potential consequence is the alteration in the number of connections into a given node. In particular, given the seemingly positive selection of editing in the high *B* cases in Fig. A.5 it might be assumed that the mechanism's ability to effectively reduce a node's *B* is all that is being selected for since fitness drops with increasing *B*. Experiments (not shown) in which the out-connection table entries are randomly re-created in the offspring indicate a significant (T-test, $p < 0.05$) drop in fitness in all cases where editing is selected for and hence evolution does appear to be shaping suitable, dynamic behaviour through the editing mechanism. Although some editing nodes are also almost certainly there due to drift/neutral processes.

Figure A.5 (right column) shows how RNA editing is selected for under all conditions when the underlying fitness landscape changes halfway through the lifecycle; an input of all 1's is applied from update cycle 50 and fitness contributions are calculated over a second NK landscape. Analysis of typical behaviour in the low *B* cases shows that one or two nodes use editing either up to or after the point of change. That

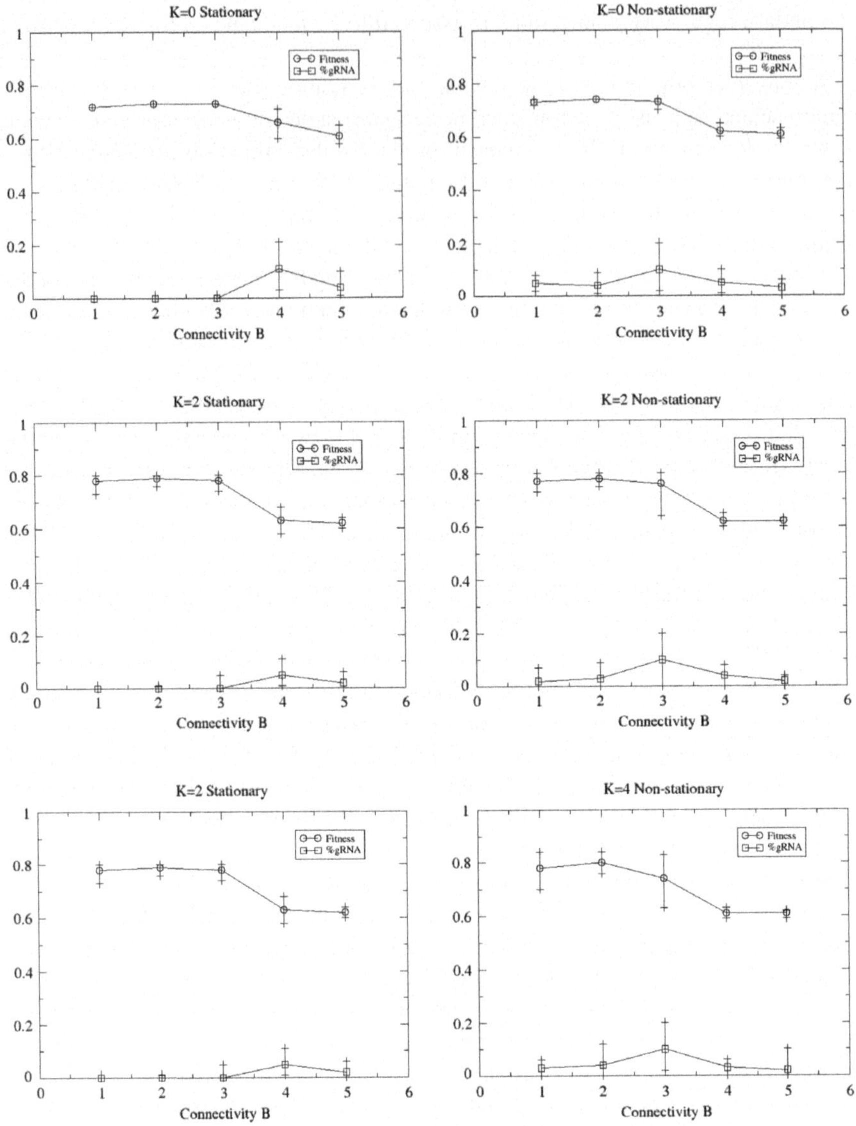

Fig. A.5 Evolutionary performance of RBN augmented with an RNA editing mechanism, after 50,000 generations, on a stationary (left) non-stationary fitness landscape (right). The percentage of nodes which use RNA editing ("%gRNA") is scaled 0-1, as is fitness is, the editing is used to make small changes to the network topology to compensate for the disruption in the environment; the RNA editing has an active (context sensitive) role in the cyclic behaviour of the networks. These general results were also found for other values of R, e.g., $R = 200$ (not shown). Thus the findings in Chap. 2 that changes in the position of optima within a fitness landscapes, in particular, promotes the evolution of larger genomes can be seen to correlate with this finding: non-stationarity promotes growth and alterations in regulatory control.

A.3 gRNA in the RBNKCS Model

Chapter 3 introduced a form the NK model which explicitly considers the effects of one species upon the others with which it has close environmental connections—the NKCS model of coupled fitness landscapes. The RBNK model is easily extended to consider the interaction between multiple GRN based on the NKCS model—the RBNKCS model. As Fig. A.6 shows, it is here assumed that the current state of the N trait nodes of one network provide input to a set of N internal nodes in each of its coupled partners, i.e., each serving as one of their B connections. Similarly, the fitness contribution of the N trait nodes considers not only the K local connections

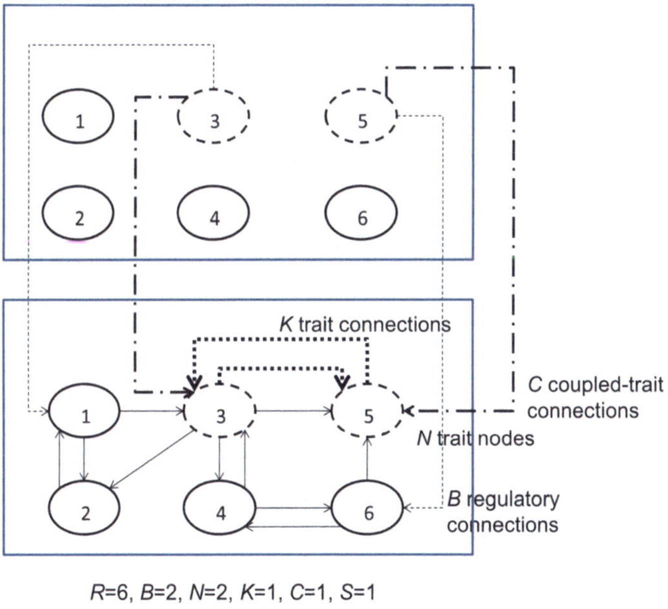

Fig. A.6 Example RBNKCS model. Connections for only one of the coupled networks shown for clarity

but also the C connections to its S coupled partners' trait nodes. The GRN update alternately.

The case of two coevolving GRN has been explored using the RBNKCS model, each evolved separately on their own NKCS fitness landscape for their N external traits. Each network updates in turn for 100 cycles. The fitness of one network is then ascertained and an evolutionary generation for that network is undertaken. The mutated network is evaluated with the same partner as the original and it becomes the parent under the same criteria as used above. Then the second species network is evaluated with that network, before a mutated form is created and evaluated against the same partner. One generation is said to have occurred when all four steps have been undertaken. Only the fitness of the species with the potential to exploit RNA editing is shown here. This general scenario is potentially of interest given the proposed role of RNA editing by cells against viruses, for example.

Figure A.7 shows how for a low degree of coupling between the two species/cells, i.e., $C = 1$, as before, low B networks result in higher fitness levels. However, regardless of K, the percentage of nodes using RNA editing increases with B. For $B = 1$ editing is not selected for on average. Recall that such networks typically exhibit a point or small attractor, and hence the coupled GRNs exist in relatively static/unchanging environments. For $B > 3$ the majority of nodes use RNA editing (>60%) but the fitness levels reached are relatively low. As above, the same experiments have been run in which the RNA editing details are randomly scrambled in offspring. Results (not shown) indicate similar high percentages of RNA editing nodes but no significant change in fitness (T-test, $p \geq 0.05$) and hence the uptake is due to drift/neutral processes within such poorly evolving systems. The exact ways in which the editing is used in the low B cases is hard to establish. Figure A.7 also shows how the same general trends occur for higher levels of coupling between the two, i.e., $C = 5$. It can be noted that some level of RNA editing is now seen when $B = 1$.

Analysis of how the percentage of RNA editing nodes varies over time finds relatively stable behaviour in the RBNK model (not shown). Figure A.8 shows example runs of how this is not the case in the RBNKCS model, rather the percentage varies over time. Indeed, it appears there is a rough correlation between periods of coevolutionary stasis with regards to fitness and a decrease in the percentage of RNA editing nodes, and vice versa, for lower values of C. Similar temporal dynamics were reported when the model was used to explore the effects of transposons [2].

A.4 gRNA in the RBNK and RBNKCS Models with Sexual Diploids

Following the findings in Chaps. 4–6, sexual diploid versions of the RBN model can be introduced. That is, an individual consists of two RBN which are each evaluated as above and their fitnesses averaged, with evolution exploiting two-step meiosis under a haploid-diploid process as described in Chap. 4. Results find the expected

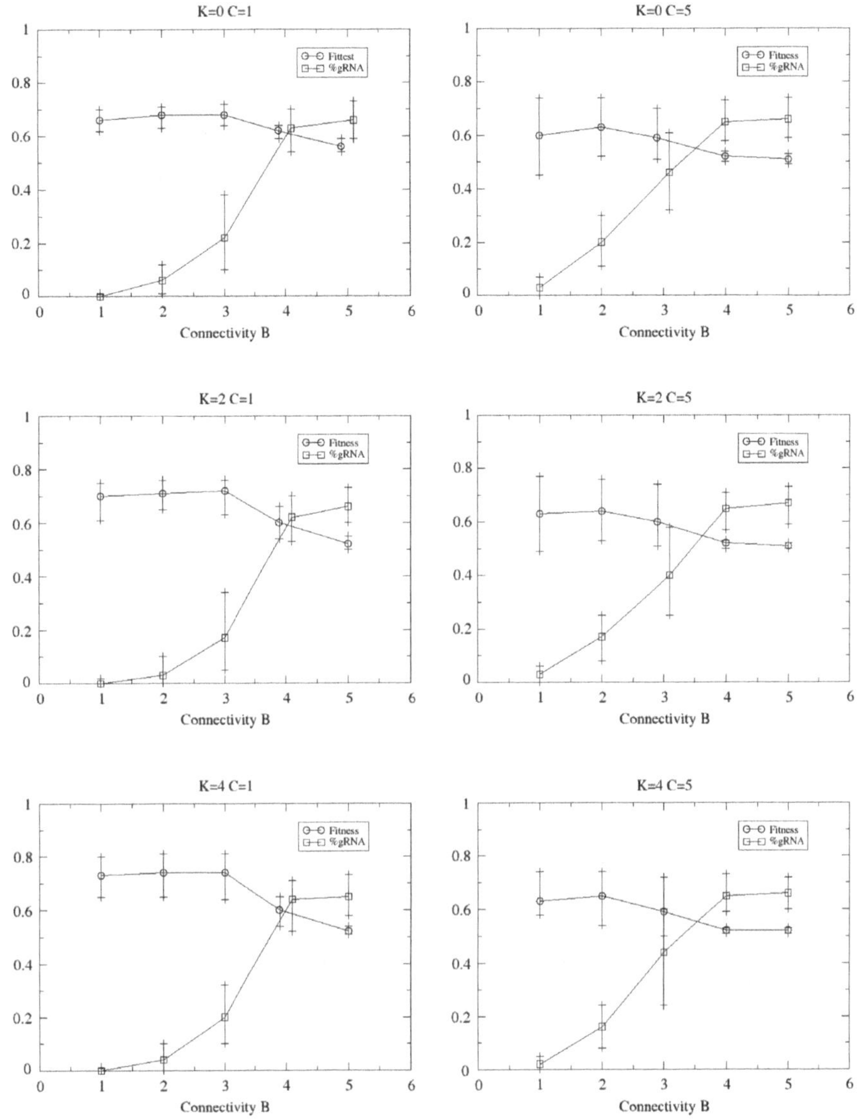

Fig. A.7 Performance of the gRNA augmented RBN coevolved against another without gRNA, after 50,000 generations, for various degrees of coupling C

increase in fitness over the asexual haploid case above for $K > 0$ (T-test, $p < 0.05$) but no significant difference in the percentages of RNA editing nodes found to emerge (T-test, $p \geq 0.05$) in both the stationary and non-stationary cases (not shown).

Figure A.9 shows examples from the coevolutionary case where a sexual diploid is paired with an asexual haploid without the RNA mechanism, i.e., as in Fig. A.7.

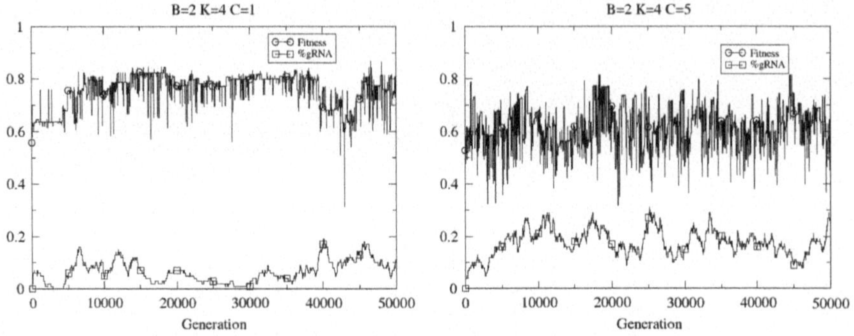

Fig. A.8 Example single runs of the coevolutionary case showing how editing is exploited to varying degrees depending upon the overall temporal dynamics of the ecosystem, rising during periods of re-adaptation and falling during periods of stasis for low coupling (*C*)

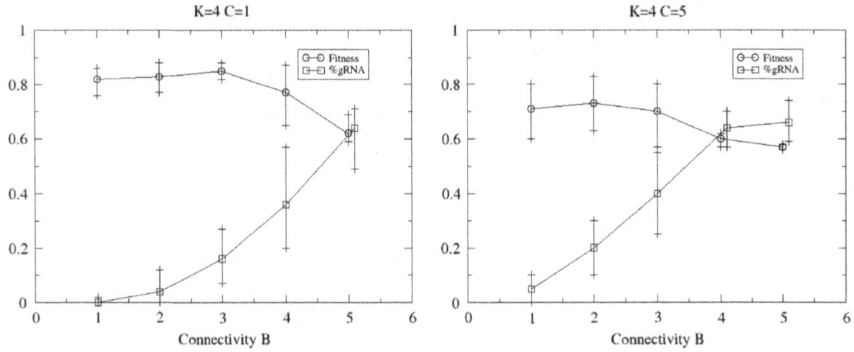

Fig. A.9 Example performance of the gRNA augmented sexual diploid RBN coevolved against an asexual haploid without gRNA, after 50,000 generations, for various degrees of coupling *C*

As can be seen, fitnesses are again improved (for $K > 0$) and the percentages of RNA editing nodes are similar to the asexual haploid case (Fig. A.7), although significantly less editing is typically seen for $B = 4$ with low C (T-test, $p < 0.05$).

References

1. Bull, L.: Evolving Boolean networks on tuneable fitness landscapes. IEEE Trans. Evol. Comput. **16**(6), 817–828 (2012)
2. Bull, L.: Evolving functional and structural dynamism in coupled Boolean networks. Artif. Life **20**(4), 441–455 (2014).
3. Bull, L.: On the evolution of Boolean networks for computation: a guide RNA mechanism. Int. J. Parallel Emergent Distrib. Syst. **31**(2), 101–113 (2016)
4. Deline, B., Greenwood, J., Clark, J., Puttick, M., Peterson, K., Donoghue, P.: Evolution of metazoan morphological disparity. Proc. Natl. Acad. Sci. USA **115**(38), E8909–E8918 (2018)

5. Derrida, B., Pomeau, Y.: Random networks of automata: a simple annealed approximation. Eur. Lett. **1**, 45–49 (1986)
6. Gray, M.: Evolutionary origin of RNA editing. Biochemistry **51**, 5235–5242 (2012)
7. Kauffman, S.A.: Metabolic stability and epigenesis in randomly constructed genetic nets. J. Theor. Biol. **22**, 437–467 (1969)
8. Kauffman, S.A.: The Origins of Order: Self-organisation and Selection in Evolution. Oxford University Press, New York, NY (1993)
9. Maas, S. (ed.): RNA Editing: Current Research and Future Trends. Caister Academic Press, Norfolk (2013)

Index

A
Allosome, 54–56, 58, 60
Asexual, 5, 18, 37, 39–46, 48, 58, 60, 71, 79, 83, 84
Attractor, 75, 82
Autosome, 54, 55

B
Baldwin effect, 3, 31–34, 36–38, 41, 44, 46–48, 56, 57, 60, 63, 71, 72, 74

C
Cambrian explosion, 63
Chloroplast, 21, 29, 47
Chromosome, 2, 47, 48, 51–60, 72, 74
Coevolution, 18, 43
Complexity catastrophe, 6, 8, 9, 15, 48

D
Differentiation, 55, 58, 63–72, 74
Diploid, 31, 33, 35–46, 48, 51–57, 59, 60, 64, 69–71, 74, 82–84
Dominance, 31, 51, 56–58, 60

E
Endomitosis, 31, 33, 35–37, 39, 47, 65, 74
Endosymbiosis, 17, 20–23, 25
Epigenetic, 63, 66–68, 70–72
Epistasis, 6, 15, 47, 48, 53
Eusociality, 63, 68–72

F
Fitness landscape, 1–3, 5, 6, 8, 9, 11–15, 17, 18, 21, 27, 29, 32, 35, 37, 39, 40, 47, 48, 51, 54–60, 63, 66, 68–74, 76, 79–82

G
Gene transfer, 5, 17, 25–29, 37
Genome, 1–3, 5–9, 11–15, 17, 18, 21, 22, 25–27, 31, 32, 35, 37, 38, 40–42, 47, 48, 51–57, 59, 60, 63–65, 67–74, 77, 79, 81
Green algae, 63

H
Haplodiploid, 68, 71
Haploid-diploid cycle, 31, 33, 35, 37, 40, 46, 47, 60, 65, 71

I
Isogamy, 31, 54, 58

K
Kin-selection, 63, 68, 69

L
Learning rate, 47

M
Mating types, 31, 51, 54, 58, 60, 74

L. Bull, *The Evolution of Complexity*, Emergence, Complexity and Computation 37,
https://doi.org/10.1007/978-3-030-40730-8

Meiosis, 2, 31, 35, 38–40, 47, 52–60, 65, 67, 69, 71, 74, 82
Mitochondria, 21, 47, 60, 71
Multicellularity, 2, 3, 63–67, 69–71, 74

N
NK model, 5–7, 11, 18, 28, 32, 35, 40, 43, 44, 48, 51, 53, 57, 63, 64, 71, 76, 77, 81
NKCS model, 17–21, 24–27, 43, 45, 75, 81

P
Ploidy, 31, 35, 36, 47

R
RBNKCS model, 81, 82
RBNK model, 75–77, 81, 82
Recombination, 5, 38–41, 46–48, 51–60, 74
RNA editing, 77–83
Ruggedness, 3, 5–10, 12, 15, 17, 19, 21–23, 28, 29, 32–34, 36–45, 47, 48, 51–53, 55–58, 60, 63–66, 68–71, 73, 74

S
Sex, 2, 3, 31, 43, 46, 47, 51, 54–56, 59, 60, 72, 74
Symbiogenesis, 17, 25, 26, 28, 29, 47, 73
Syngamy, 2, 31, 33, 35–39, 47, 60, 74

T
Termites, 68, 71

V
Volvox, 63

X
X0 system, 51, 58–60
XY system, 51, 58–60

Z
Z0 system, 58
ZW system, 58, 60

The manufacturer's authorised representative in the EU is Springer Nature Customer Service Centre GmbH, Europaplatz 3, 69115 Heidelberg, Germany. If you have any concerns regarding our products, please contact ProductSafety@springernature.com

Printed and bound by CPI Group (UK) Ltd, Croydon, CR0 4YY
30/09/2026
02225607-0001